COMPUTERS, BRAINS AND MINDS

AUSTRALASIAN STUDIES IN HISTORY AND PHILOSOPHY OF SCIENCE

VOLUME 7

COMPUTERS, BRAINS AND MINDS

Essays in Cognitive Science

Edited by

PETER SLEZAK

Centre for Cognitive Science, University of New South Wales

and

W. R. ALBURY

School of Science and Technology Studies, University of New South Wales

KLUWER ACADEMIC PUBLISHERS

DORDRECHT / BOSTON / LONDON

Library of Congress Cataloging in Publication Data CIP

Computers, brains, and minds : essays in cognitive science / edited by Peter Slezak and W.R. Albury.
p. cm. -- (Australasian studies in history and philosophy of science ; v. 7)
Papers from a special symposium organized for the joint conference of the Australasian Association for the History, Philosophy, and Social Studies of Science and the Australasian Association of Philosophy held in late 1985 at the University of New South Wales in Sydney, Australia.
Includes index.
ISBN 9027727597 (hardback)
1. Cognitive psychology--Congresses. 2. Human information processing--Congresses. 3. Neuropsychology--Congresses. 4. Cognition--Congresses. I. Slezak, Peter, 1947- . II. Albury, W. R. III. Australasian Association for the History, Philosophy, and Social Studies of Science. IV. Australasian Association of Philosophy. V. Series.
BF311.C59 1988
153--dc19 88-14698
CIP

Published by Kluwer Academic Publishers,
P.O. Box 17, 3300 AA Dordrecht, The Netherlands

Kluwer Academic Publishers incorporates
the publishing programmes of
D. Reidel, Martinus Nijhoff, Dr W. Junk and MTP Pres

Sold and distributed in the U.S.A. and Canada
by Kluwer Academic Publishers,
101 Philip Drive, Norwell, MA 02061, U.S.A.

In all other countries, sold and distributed
by Kluwer Academic Publishers Group,
P.O. Box 322, 3300 AH Dordrecht, The Netherlands

Printed in The Netherlands

on acid free paper

TABLE OF CONTENTS

FOREWORD

The institutionalization of History and Philosophy of Science as a distinct field of scholarly endeavour began comparatively early – though not always under that name – in the Australasian region. An initial lecturing appointment was made at the University of Melbourne immediately after the Second World War, in 1946, and other appointments followed as the subject underwent an expansion during the 1950s and 1960s similar to that which took place in other parts of the world. Today there are major Departments at the University of Melbourne, the University of New South Wales and the University of Wollongong, and smaller groups active in many other parts of Australia and in New Zealand.

"Australasian Studies in History and Philosophy of Science" aims to provide a distinctive publication outlet for Australian and New Zealand scholars working in the general area of history, philosophy and social studies of science. Each volume comprises a group of essays on a connected theme, edited by an Australian or a New Zealander with special expertise in that particular area. Papers address general issues, however, rather than local ones; parochial topics are avoided. Furthermore, though in each volume a majority of the contributors is from Australia or New Zealand, contributions from elsewhere are by no means ruled out. Quite the reverse, in fact - they are actively encouraged wherever appropriate to the balance of the volume in question.

R. W. HOME
General Editor
Australasian Studies in History
and Philosophy of Science

PREFACE*

The present volume emerges from an attempt to bring together such interest in cognitive science as there was in Australia and New Zealand in late 1985. The occasion was a special symposium organised for the joint conference of the Australasian Association for History, Philosophy and Social Studies of Science and the Australasian Association of Philosophy, held at the University of New South Wales in Sydney, Australia. Remote though Australia may be physically from the principal centres of activity in this area, in terms of Dennett's logical geography,[1] it remains to be determined where *Terra Australis* should appear in its relation both to the 'East Pole' of MIT High Church Computationalism and to the 'Zen Holism' of the American West Coast. Contrary to the flat earth appearance of Dennett's map, the contributions to this volume suggest that going far enough West brings the traveller back to the East Pole, with the antipodes standing nearer to the mother church than might be thought from mere compass direction. Be that as it may, it was the first visit to our far flung shores of the reigning Pope of Computationalism, Jerry Fodor, which was the occasion for a revitalisation of the faith (though, of course, there are well-known exegetical grounds for holding that Australian philosophers were the earliest founders of the religion, in the form of 'old testament' materialism as expounded by Place, Smart and Armstrong). Thus, Fodor's visit to give the keynote address to the Australasian cognitive science symposium was a timely event, perhaps even more a 'revival' meeting than a papal appearance, since it was also the occasion for announcing the establishment of the first graduate degree program in cognitive science in Australia at the University of New South Wales.

Given the origins of this volume in the 1985 symposium, most of the papers appearing in it have an Australian connection, with some of them, like Fodor's, actually presented for the first time at the symposium itself. Others, like Michael Arbib's and Bernard Berofsky's, although not presented at the symposium, were contributed at the invitation of the editors. Though not entirely by design, it is significant that the papers collected here constitute a representative survey of contemporary concerns and debates surrounding cognitive science. We believe that the volume will contribute to these discussions while also stimulating further interest in these exciting developments in Australia and New Zealand, wherever this antipodean region might ultimately be located in the logical cartography of the field.[2]

PETER SLEZAK and W. R. ALBURY

NOTES

* The editors are very grateful to Anne Warburton and Anita Soekarno for their assistance in the preparation of this volume. The support of Apple Computer Australia is also gratefully acknowledged.

[1] D. C. Dennett (1986) 'The Logical Geography of Computational Approaches: A View from the East Pole', *The Representation of Knowledge and Belief*, M. Brand, and R.M. Harnish, (eds.), University of Arizona Press.

[2] In addition to the present volume, a further set of papers arising from the 1985 Cognitive Science Symposium has been published in W.R. Albury and Peter Slezak (eds.) (1988) *Dimensions of Cognitive Science*, Centre for Cognitive Science (University of New South Wales).

PETER SLEZAK

INTRODUCTION

While the papers in this volume were written independently of one another, there is a clear overall unity in their concerns, with a few common threads discernable throughout. These reflect issues of central interest within cognitive science and include in particular the computational view of the mind. The purpose of this introductory essay is to act as a guide to the contributions by summarising their central arguments and also offering some brief commentary where appropriate.

1. J. A. FODOR: 'WHY THERE *STILL* HAS TO BE A LANGUAGE OF THOUGHT'

As the title suggests, Jerry Fodor is returning to defend the thesis of his landmark book *The Language of Thought*.[1] In this book Fodor provided the invaluable philosophical service of articulating clearly the sense in which mental or cognitive processes are to be seen as computational. For all the loose talk of the 'computer metaphor' for the mind, Fodor spelled out in detail the reasons for taking this view seriously and literally – essentially because all our most plausible psychological theories are committed to construing cognitive processes as computational. That is, Fodor pointed out that, whatever their details, the general structure of theories of cognition presuppose underlying computational processes and, specifically, a representational system in which these processes are carried out. Theories of deciding, perceiving, learning, language acquisition, all presuppose that the organism has available a medium of internal representation of great richness and this representational system must have properties similar to natural languages. In this sense Fodor argued for the existence of a 'private language' though one not subject to Wittgensteinian concerns. For as Fodor noted,[2] whatever Wittgenstein is supposed to have shown to be impossible, Fodor's theory posits something quite different and more akin to the machine language of a computer. After all, he noted "there *are* such things as computers and whatever is actual is possible". Thus Fodor relied heavily on the machine analogy in a number of respects which were explicitly spelled out. He wrote:

Peter Slezak and W.R. Albury (eds.), Computers, Brains and Minds, 1–21

When we think of an organism as a computer, we attempt to assign formulae in the vocabulary of a psychological theory to physical states of the organism (e.g., to states of its nervous system). Ideally, the assignment should be carried through in such a fashion that (some, at least) of the sequences of states that are causally implicated in the production of behaviour can be interpreted as computations which have appropriate descriptions of behaviour as their 'last line'. The idea is that, in the case of organisms as in the case of real computers, if we get the right way of assigning formulae to the states it will be feasible to interpret the sequence of events that *causes* the output as a computational *derivation* of the output. In short, the organic events which we accept as implicated in the etiology of behaviour will turn out to have two theoretically relevant descriptions if things turn out right: a physical description by virtue of which they fall under causal laws and a psychological description by virtue of which they constitute steps in the computation from the stimulus to the response.[3]

In this attempt to capture the appropriate level of theory at which the computational states and relations can be expressed, of particular importance for Fodor is ordinary belief-desire talk or the propositional attitude talk of 'folk psychology' which will constitute by and large the correct level. That is, believing, desiring, fearing, hoping, intending, learning, perceiving, etc., will be explicated in terms of some corresponding computational relation to a formula of the internal code. Fodor has thus been an advocate of the indispensability of commonsense psychological explanation, specifically because of its implicit commitment to the intentional contents of mental states and their causal role in the aetiology of behaviour and other mental states. This is what Fodor refers to in his present paper as 'intentional realism' and it is important to note in passing that not all philosophers share this realist view of the propositional attitudes. Dennett, Stich and the Churchlands, for example, are skeptical about this matter for a variety of different reasons.

Fodor's claim that there is a language of thought, however, goes significantly beyond the mere commitment to intentional realism. That is, the language of thought (LOT) thesis goes beyond the claim that there are internally represented mental states which have a causal role in behaviour to the further claim that these internal states also have a complex constituent structure which is appropriate to their semantic content through mirroring the structure of their associated objects.

The clarification and defence of this thesis with its 'classical' computational conception of the mind has become particularly urgent in recent times with the dissent growing from the direction of 'new connectionism'. Fodor's view can been seen as favouring the Turing/von Neumann computer architecture as appropriate to explaining the structure of the mind, and the arguments here have recently been redeployed against the claims of 'connectionism'.[4] The claim of LOT for structured mental states is that beyond having a causal role and being semantically evaluable, the mental states must also have sub-parts which are themselves semantically evaluable. These sub-formulae will be the atomic components making up complex formulae and will provide the explanation for the commonalities among different molecular formulae which would appear accidental and utterly inexplicable on a non-compositional account. Fodor has called this property of mental representations their 'systematicity', which refers to the idea that the ability to produce or understand a sentence is intrinsically connected to the ability to produce or understand certain others – an ability which can only be accounted for on the assumption of a combinatorial or constituent structure for mental representations. In this respect, the systematicity of thought is exactly like the systematicity of language and explains the connection between our ability to think certain thoughts and our ability to think certain other related ones. The intentional realist's mere postulation of semantically evaluable internal states (perhaps neurological states) does not go far enough, since being in one such state need be in no way connected with being in another semantically related one. Fodor's argument here for a language of thought rests on the need to explain the remarkable contingent fact of the existence of connections among thoughts. In Fodor's view, the overall situation is quite simple: the need to explain such central phenomena means that "As things stand now, the cost of not having a Language of Thought is not having a theory of thinking".

2. D. H. MELLOR: 'HOW MUCH OF THE MIND IS A COMPUTER?'

It is on this last point that Mellor's paper provides a valuable contrast, for he argues precisely against the broad generality of a computational view of mind. Specifically, Mellor's thesis is that, while the limited domains of perception and inference which involve belief may be computational, the rest of the mind, and therefore most of it, is not. Details aside, as a general methodological matter, Mellor's view has the disadvantage of complicating the story of the mind through requiring radical discontinuities between the mechanisms subserving belief and everything else. *Ceteris paribus*, we would expect uniformity in the fundamental processes underlying cognition. In this respect Mellor's view confronts a difficulty not faced by the more thoroughgoing rejection of Fodor's computationalism to be seen in Patricia Churchland.[5] Churchland rejects the entire 'sententialist' approach as a model of cognition, tied as it is to the concepts of language and folk psychology, whereas Mellor concedes its application at least for truth bearing states to be found in perception, inference and belief.

In agreeing with the need to postulate propositional attitudes, Mellor avows an intentional realism, but it is precisely in the degree to which he refrains from going further that he departs from Fodor and the LOT view. Fodor's computational account turns on construing the propositional attitudes as having (a) a semantically evaluable propositional content and (b) an attitude such as belief or desire which is taken to be a computational relation between the organism and the proposition. Mellor does not accept the generality of this view and accepts this analysis for only those propositional attitudes involving belief in some way. Mellor articulates an account here in which he emphasizes the syntactic basis of computation and its underlying causal processes turning on specific intrinsic properties of the tokens involved. This is what Fodor has called the 'formality condition'.[6] However, Mellor presses the question: 'When is a causal process a computation?' Mellor's answer is that the criterion for a causal function's being syntactic depends on semantics. In this sense, Mellor argues that a purely formal syntax is not autonomous, but presupposes a semantics and, in this sense, the semantics does not merely exploit syntactic properties but actually creates them. Thus, Mellor suggests that it is only when we know where the semantics comes from that we will know what it takes to be a computer and, in particular, how much of the mind may be one.

In the case of computers, Mellor points out that it is obvious the semantics derive from us, but it has been precisely the asymmetry in this regard which has been seen as the fundamental problem for understanding intentionality, content and how the mind might be seen as computational. In particular, the apparent discrepancies between folk-psychological type-individuation of mental states (so-called 'wide psychology') and the purely syntactic or formal individuation ('narrow psychology') has been a source of on-going perplexity and debate.[7]

For Mellor it is essential to distinguish those causal processes which are truly syntactic and computational through having a semantics, from those which are merely non-representational processes which might nevertheless permit an algorithmic description. Thus, Mellor is relying on a distinction which is familiar from debates notorious in connection with grammars and their 'psychological reality'. In this context, too, a distinction has been pressed claiming that a system may be describable in terms of rules which nevertheless it may not be said actually to follow. Here Mellor argues that mental processes other than perhaps perceiving and inferring are not computational because the states involved do not represent anything in the required sense. Thus, sensations and pains, for example, do not represent their causes in the way that a belief might and are more like the case of a mass which merely functions in accordance with Newton's laws but cannot be said to use them to work out a resultant force. It is here that Mellor confronts Fodor's LOT claims most directly. He takes an example which happens to be a paradigm case for Fodor and, indeed, precisely one used in the original book[8] to argue the case *for* the inescapability of a computational account. Relying on the distinction just noted, Mellor suggests that decision theoretical accounts of action in terms of some calculation of expected utilities etc., are plainly false. For, he suggests, although one *might* conceivably act through performing such a decision-theoretic calculation, nevertheless one plainly does not. Mellor even concedes that such a model may accurately capture how beliefs and desires cause me to act, but he says "That doesn't mean I compute my action from them, and I don't". Here Mellor seems to be relying on some distinction between what I am aware of doing and what is unavailable to consciousness, but this hardly seems adequate to deny the claim that the underlying causal processes are computational, albeit unconscious. Reliance upon the comparison with a mass obeying Newton's laws seems particularly questionable in the present case, where we do, after all, have good reason to attribute the causes of action to internal representations. In this case it is not at all clear what could be meant by saying that we simply react in the way a mass does rather than act as a consequence of computational processes. It is worth noting that Mellor's position here relies on the kind of anti-realism adopted by the critics of Chomsky in relation to the status of the formalisms of competence theories in linguistics. In response, Chomsky's repeated question has been to ask why we should not take the implications of our best theories seriously and literally when they make certain posits, just as we ordinarily do elsewhere in science.[9] Of course, this recalls Fodor's essentially similar strategy in the original argument for a computational view of LOT,[10] where he argued that we are provisionally committed to a computational view insofar as we take our best available theories seriously.

Aside from an instrumentalism regarding the status of formal models, Mellor's argument depends principally on his claim that, besides belief, the other propositional attitudes do not embody truth-bearing propositions. Hoping, fearing or wanting does not represent some fact as obtaining. Notice that Mellor is making a much stronger claim than merely adverting to the content-bearing character of propositional attitudes, for it is not just the semantic evaluability he points to as relevant. Mellor appears to require that the mental states be not only semantically evaluable, but assertive in the sense of actually purporting to be true. Commands and questions, for example, fail on this score since, Mellor claims, they don't embody a proposition representing some fact as obtaining. Mellor's argument here seems highly counterintuitive, for the mere having of representational content and thereby truth conditions is not to assert their obtaining. Nevertheless, it is this failure which Mellor takes to disqualify the propositional attitudes other than belief from embodying information and being computational. Even if Mellor is right on this point, it may not impugn a computational account of mind *per se*, but only this particular view of what it is to be computational. There are, of course, quite independent grounds for construing the information processing of the mind as computational at a level more remote from the propositional attitudes of folk psychology and closer to the activities of the neurones as, for example, in the case of the retina or the regular architecture of the cerebellum, to cite only two well-known examples in which the computational processes are becoming reasonably well understood.

3. KIM STERELNY: 'COMPUTATIONAL FUNCTIONAL PSYCHOLOGY'

Sterelny too takes up the computationalist thesis, considering it together with the closely allied notion of homuncular functionalism. The latter advocated by Dennett as a kind of modularity is usefully contrasted with Fodor's recent, though quite different, modularity claims.[11] Sterelny introduces grounds for skepticism about the thoroughgoing functionalist idea that there is a purely abstract, autonomous level of description independent of neural or other realization. He suggests that some functional states, like sexual desire, may well be more intimately coupled with their biological realizations. The advantage of homuncular functionalism over previous unitary accounts is in its providing for multiple levels of psychological description.

Sterelny points out that the idea of cognition as computation has been closely connected with functionalism. Indeed, perhaps more closely than he suggests for, in fact, the earliest version of functionalism due to Putnam was stated in terms of Turing machines and in this form gave a precise sense to the idea of the abstractly specified states of a device, independent of its realizations. Sterelny enumerates many of the virtues of the computational point of view, noting in particular a point we have seen raised by Fodor and Mellor, namely, the way in which causally related states subserve the information processing transformations. Sterelny points to the fact that the information processing functions performed by our brains have no access to distal causes, but only to intrinsic properties such as the neuronal signalling patterns. This is the problem which has continued to cause great consternation among philosophers, since intentional states appear to require going beyond purely formal or syntactic properties and adverting to semantic properties such as truth and reference. Sterelny's way of putting the point here in terms of the brain's exclusive access to intrinistic, formal properties is a useful antidote to prevailing skepticism about the adequacy of a purely syntactic or 'narrow' psychology to be seen in Curge[12] and others. Certainly when the formal processes of a computer are considered, their interpretation or meaning derived from our own design and purposes leaves obscure how mental states in us could have intrinsic content. Nevertheless, it would seem that whatever the shortcomings of the computational view in other respects, this 'formality condition' or 'methodological solipsism' ought not to be among them, since neuroscience must be just as formal and solipsistic in this regard. However we are to make sense of the semantic, intentional states of minds, here at least there is a symmetry with computers.

The motivation for a 'wide' psychology to accomodate semantic properties of intentional states comes largely from folk psychology. In the course of considering the dispute about the role of folk psychology in cognitive science, Sterelny notes that folk psychology is semantic psychology in the sense of positing states with contents about the world. As just observed, it is this which leads to the apparent conflict between folk psychology and the computational point of view. However, Sterelny point out that whatever the ultimate fate of folk psychology, it need not threaten homuncular functionalism which posits a hierarchy of levels in the analysis of mind. As far as the computational point of view is concerned, Sterelny draws the moral of the conflicting perspectives in asking what would be left of the view that cognition is symbol processing if semantic properties do not find a place in a mature cognitive science.

Sterelny goes on to raise the question of computationalism in the form of a concern about its biological plausibility. Among the central issues here is the claim by Fodor originally in his *Language of Thought*[13] concerning the process of concept acquisition. On the face of it, Fodor's claim is startling and implausible. Baldly stated, it is the view that we can only acquire those concepts already represented in our innate language of thought and in this fundamental sense we can never really learn new concepts. Patricia Churchland[14] has drawn out the extremely counter-intuitive consequences of Fodor's view, and in taking up these criticisms here, Sterelny considers possible avenues of salvation for Fodor's thesis via the notion of 'triggering' which can render innateness claims more subtle and correspondingly more plausible. A further basis for Churchland's skepticism about Fodor's claim for a language of thought is what she dubs the 'infralinguistic catastrophe'. This is a reference to the fact that there are intimate continuities between human language users and non-linguistic animals or pre-linguistic children. This suggests the implausibility of any account of cognition which is too closely modelled on the properties of language. It is suggested that this is particularly so in view of the late arrival of language in the evolutionary process. In discussing this problem, Sterelny makes the elegant point that this argument can be turned on its head, for it is likely that language should have inherited fundamental properties of the systems from which it must have evolved. Nevertheless, overall Sterelny is sympathetic to the biologically based criticisms of computationalism and concludes that while this may well be wrong, the functional theory is more certain to be vindicated.

4. BERNARD BEROFSKY: 'BELIEF AND RESPONSIBILITY'.

Bernard Berofsky takes up central problems we have seen raised in the preceding discussions concerning the relation of folk psychology to a prospective cognitive science. In particular, he considers the paradigmatic propositional attitude *belief*, which has a particular interest beyond its centrality in the foregoing regard – namely, its essential role in ascriptions of responsibility. Berofsky's discussion illustrates the way in which the problems at the center of debate in cognitive science have broader implications. Specifically, the problem already noted concerning the semantic content of intentional states such as belief threatens the notion which is a condition of moral responsibility. That is, if beliefs are individuated 'widely' by content and not individualistically or 'solipsistically', then it seems they cannot be construed as guiding behaviour. The concern here has a formal parallel with the traditional worries about responsibility deriving from determinism in so far as a necessary condition for attributions of responsibility is threatened in both cases: autonomous agency in the case of determinism, and the related notion of belief-guided behaviour in the present case. Berofsky's well-known contributions to the literature on determinism make his present discussion particularly interesting by showing a link between earlier philosophical debates and the most recent ones within cognitive science. Although the question of responsibility provides the motivation for Berofsky's discussion, the puzzles he discusses have independent status and interest. Among the troubling problems for belief canvassed is the difficulty arising from the potentially infinite set of our beliefs and how this fact is to be reconciled with our finite neurological states; there is the further threat to the idea of a belief as an internal state derived from the social nature of belief and the normative assumptions of rationality which seem to be imposed upon our belief attributions. Berofsky provides a valuable discussion of the difficulty for belief posed by Dennett's instrumentalism based on the possibility of 'emergence' in systems where there may be no beliefs explicitly tokened. Implicated in this general concern to account for beliefs is what Berofsky refers to as 'sententialism', namely the view that belief states and their role in the aetiology of behaviour are to be explained in terms of 'mental sentences'. After considering a variety of potential difficulties for a strong version of this claim, Berofsky argues that a weaker sententialism is compatible with the existence of other representational forms and may be adequate for preserving the notion of belief required for responsibility attribution. Even more extreme scenarios such as that envisioned by Paul Churchland permit saving an adequate notion of sententialism, leaving only extreme eliminativists and 'rabid anti-representationalists' as unvanquished. While Berofsky rightly observes that there can be no *a priori* guarantees that cognitive psychology will always underwrite moral agency, his thorough exploration of the current issues leads to his conclusion that there is no need to worry since there is no case against belief.

5. CHRIS MORTENSEN: 'MENTAL IMAGES: SHOULD COGNITIVE SCIENCE LEARN FROM NEUROPHYSIOLOGY?'

Until very recently, the details of neurophysiology had not figured prominently in philosophical discussions of the mind. Even now the Churchlands are something of an anomaly with their 'neurophilosophy', advocating profound reorientation of cognitive science and specifically challenging the Fodorian computational, 'sentential' view we have seen. On the face of it, the neglect of neuroscience has been odd if only because modern materialism explicitly identified the mind with the brain. But for various reasons the actual details were taken to be irrelevant, and for a long time token gestures sufficed in the form of passing reference to the firing of 'C-fibres'. The computationalist approach central to current views of mind reinforces this neglect of the brain on the grounds that neuroscience is concerned with mere implementation details and, consequently, is at the wrong level to capture the generalisations of interest concerning cognition. This functionalist, top-down stratagem dominates philosophical discussions and inhibits interest in the details of neuroscience at a time when, paradoxically, the substantive details of other disciplines have been receiving serious attention by philosophers in the spirit of naturalized epistemology.

Contrary to this prevailing view, we have three papers in the present collection which advocate taking neuroscience seriously as having relevance for philosophical questions. The first of these is Mortensen's article which cautions against what he calls the 'AI fallacy' which assumes too strong a relevance of artificial intelligence models to biologically realisable systems. The argument for attending to the constraints derived from evolved, biological systems is offered here specifically in connection with the perennial problem of mental imagery. In the completely different domain of action theory Slezak makes a similar point, arguing for the pointlessness of traditional philosophical approaches based on naive folk theories, particularly when there is a wealth of relevant empirical material already available from neuroscience bearing on central philosophical questions. In the same vein, Arbib's paper takes detailed work in neurophysiology as its point of departure and criticizes Fodor's recent arguments for delineating the 'modularity of the mind'. These papers reflect what is now a significant trend away from purely top-down approaches – a trend seen even within AI itself, as the neurologically inspired models of the 'new connectionism'[15] come to dominate discussions.

Although the topic of Mortensen's paper, mental imagery, is one with a long history, it has recently become one of the centrally debated topics within cognitive science, perhaps largely as a consequence of the striking research results obtained by Shepard[16] and his colleagues. These results appear to warrant taking mental images seriously as some kind of iconic or pictorial form of internal representation. Thus we see Block as editor of a recent collection of papers on imagery reflecting the current sense of novelty and breakthrough where he writes:

> After fifty years of neglect during the heyday of behaviourism, mental imagery is once again a topic of research in psychology. Indeed, with the emergence of a truly spectacular body of experiments, imagery is one of the hottest topics in cognitive science.[17]

In a nutshell, the recent debate has concerned the question of whether mental representations which subserve imagery have properties making them more akin to pictures or more like sentences.

Before turning directly to Mortensen's paper, it is worth pausing to make some historical reflections on the issues raised here. Amid the self-congratulatory hyperbole which abounds in these heady days of the cognitive revolution, there is a sobering lesson to be learned from re-examining some anticipations of the current debates. For example, it is striking to notice exactly the same issues discussed by Descartes, who moreover, already offered a solution along the lines of recent accounts specifically to avoid the pitfalls inherent in the pictorialist view. As Descartes was well aware, the theory which is intuitively or introspectively most appealing is the pictorial or iconic one, in spite of the conceptual and empirical difficulties in making sense of it. It is clearly through appearing to provide support for the intuitive account that the results of Shepard are surprising. But despite its appeal, Descartes explains why any literal postulation of images is misguided, and he does so in essentially the same terms as are found in contemporary discussions. Though neglected by philosophers in the Analytic tradition until recently, the corpus of Descartes's scientific writings reveals a different philosopher entirely from the one familiar to students of the *Discourse*. Studied in isolation from the scientific works *Dioptrics, Meteorology* and *Geometry* to which it was a preface, the *Discourse*, just like the *Medita-*

tions, provides a distorted picture of Descartes which looks, perhaps predictably, something like an Anglo-American analytic philosopher. A more balanced view of Descartes which accommodates all his writings has recently been emerging[18] and justifies seeing Descartes in a profoundly different light, as Clarke[19] has recently described him: namely, "a practising scientist who, somewhat unfortunately, wrote a few short and relatively unimportant philosophical essays". There is extreme irony in the fact that Descartes, who is blamed for foisting dualism upon subsequent generations, was the very paradigm of a contemporary naturalized philosopher and cognitive scientist and, moreover, one for whom the idea of 'neurophilosophy' in particular was embodied in his practice. Of particular relevance here, however, is the point that to anyone familiar with Descartes's writings in his *Dioptrics*, the contemporary literature on imagery produces a strong sense of *déjà vu*. Remarkably, Descartes is fully aware of the problem of the homunculus in discussions of imagery and he avoids it through an explicit 'information processing' account. In particular this account permits treating imagery in terms of represention s which are abstract and arbitrary in the appropriate sense of being symbolically coded. Descartes employs the analogy of the blind man's cane which provides detailed information about the textures of objects, but does not literally convey any substance along its length. In the same manner he proposes to explain the way in which the eyes convey information about the external world.

Descartes's remarks in the *Dioptrics* illustrate his remarkable anticipations of present discussions. In view of the currency of the doctrine criticized by Descartes, and in view of the general neglect of this dimension of Descartes's work, it is interesting to read his comments at length:

(. . .) you must beware of assuming, as philosophers ordinarily do, that it is necesary for sensation that the soul should contemplate certain images transmitted by objects to the brain; or at any rate you must conceive the nature of these images quite differently from their way of thinking. For since they have no notion of images except that they must be like the objects they represent, they cannot possibly explain how they can be produced by these objects, and received by the external sense organs, and transmitted by the nerves to the brain.

(. . .) Now we must hold a quite similar view of the images produced on our brain; we must observe that the problem is to know how they can enable the soul to have sensations of all the various qualities in the objects to which the images refer; not, how they can resemble the objects.

(. . .) And when it is transmitted to the inside of our head, the picture still retains some degree of its resemblance to the objects from which it originates. But we must not think that it is by means of this resemblance that the picture makes us aware of objects – as though we had another pair of eyes to see it, inside our brain; I have several times made this point; rather, we must hold that the movements by which the image is formed act directly on our soul *qua* united to the body, and are ordained by Nature to give it such sensations.[20]

Mortensen's discussion is in this Cartesian spirit which repudiates a literally pictorial or iconic notion of imagery. He suggests that certain propositional attitudes are best construed as having contents which are not verbally describable. He summarizes the advantages of such non-verbal beliefs, such as the capacity to give a unified account of human and pre-linguistic species in view of their evolutionary continuity. This is the kind of charge which Patricia Churchland had made against Fodor's LOT thesis, but it requires great care, since it is surely a straw-man to construe the LOT thesis as attributing a commitment to natural language as the basis for cognition in dumb animals. On the other hand, the top-down computational approach does seem to promote excessive anthropocentrism, making it difficult to see how the requisite evolutionary continuities down to Aplysia will be preserved.

In discussing the debate between 'iconophile' Kosslyn and 'iconophobe' Pylyshyn, Mortensen draws attention to the point we have seen made by Descartes: namely, that the capacity for representations to subserve imagery resides in their preserving all the necessary information processes and not in any literal pictorial resemblance. More generally, Mortensen cautions against ignoring the constraints inherent in biologically evolved systems which must provide insights into the phenomena of greatest interest to cognitive science. Adopting an eclectic view, Mortensen concludes that, in the end, as an empirical matter regarding the neurological substrate, organisms might exploit spatial properties of this evolved medium to capture relevant geometrical isomorphisms. This is an important empirical issue, though there is an appearance of question-begging in his asking: "Prior to language, how *otherwise* might an organism go about representing information about shapes within itself?".

6. PETER SLEZAK: 'HOW *NOT* TO NATURALIZE THE THEORY OF ACTION'

Slezak's discussion provides further support for turning to neuroscience as a source of insight, if not answers to philosophical problems – in this case in the area of action theory. The problem of action is another long-standing philosophical puzzle and Slezak's strategy is to undertake a close examination of Myles Brand's recent book *Intending and Acting*[21] which is of particular interest as a vehicle for exploring questions central to contemporary cognitive science. Philosophical action theory presents an anomaly through having apparently been left behind in the cognitive revolution. Slezak observes that whereas the philosophy of mind has become thoroughly naturalized, the closely allied philosophy of action has continued along its former path, wending its way between the putative insights of folk psychology and conceptual analysis. Not just in its method, but in its content too, philosophical action theory is little different from the age-old puzzle of free will, albeit in a slightly different guise. Thus it is that Brand's book derives its interest as a focus for discussion, for on the one hand Brand announces his intention to bring to action theory a new, naturalized phase. On the other hand, however, Slezak argues that Brand fails in this attempt and manages only to restate a traditional *a priori* commitment to volitions or 'conation' as a necessary component of action; in short, Brand attempts to justify some notion of the will. Brand's work is interesting because it deals with various theories of action in AI and information processing psychology including the seminal TOTE model of Miller, Galanter and Pribram.[22] Slezak's detailed analysis of Brand's approach to these matters is offered as an instructive case study of such broader issues as the place of folk psychology in cognitive science and the character of a naturalized philosophy as well as the specific question of the causal antecedents of action. It is on this latter issue that Slezak draws attention to the anomaly in professing a naturalized action theory which utterly neglects the vast neuroscience literature concerning motor control. Whatever else it may be, the problem of motor control in neuroscience *is* the problem of the causal antecedents of action. Moreover, our understanding here is vast, detailed and firmly grounded in experiment. Furthermore, Slezak makes the point that such apparently bottom-up approaches cannot be dismissed with the functionalist stratagem of regarding them as mere implementation detail. Insofar as such neuroscience can be said to be botom-up, this reflects more a preference for working within the constraints imposed by the embodying medium than a functionally irrelevant, mere realization detail. That is, studies of motor control illustrate the general point that neuroscience is frequently and inherently as much top-down as bottom-up. Thus, it is in the course of his critique of Brand that Slezak introduces illustrative materials from neuroscience which he suggests provide insights into the causal antecedents of action and whici, moreover, readily demonstrate the pointlessness of a philosophical recidivism which takes intuitions informed only by folk psychology as an infallible guide to systematic theory. Thus, independently of their specific bearing on theories of action, Slezak's arguments are offered in support of the kind of general skepticism concerning folk psychology articulated by the Churchlands, though in this case, in the domain of motor control rather than in the more usual domain of cognition.

7. PHILIP CAM: 'NOTES TOWARD A FACULTY THEORY OF COGNITIVE CONSCIOUSNESS'

Cam's paper is a further illustration of the empirically oriented approach to the philosophy of mind and is concerned to sketch the outlines of a faculty model of mental processes. Cam takes a faculty to be a more or less self-contained processor characterised in part by the format of its representations, the kinds of operations performed over these representations and its connections to other faculties, sensory systems and effector systems. In particular, Cam argues that the faculties have a special role in conscious states which he identifies with the operations of the various faculties themselves. Cam begins by contrasting his particular conception with the faculty theories of Fodor[23] and Gardner,[24] specifically dissenting from Fodor's non-modular, 'holistic' view of higher cognitive functions. Cam proposes modules which are more global, being responsible for both perception and cognition. Among the evidence cited in support of his faculty model, Cam mentions the kinds of disconnections or dissociations found in experiments by Gazzaniga based on lateralization of functions between hemispheres. In addition, Cam examines the mental rotation experiments of Shepard from a phenomenological point of view, proposing an alternative analysis consistent with his faculty scheme. The suggestion that there are different representational formats makes contact with the issues we have seen raised in Mortensen's paper concerning the debate between descriptional and pictorial accounts of imagery. Cam takes up the consideration of further experimental work by Nisbett and Wilson which bears on the unreliability of subjective, introspective reports and suggests an alternative explanation to their intellectualist, unconscious inference account.

Cam cites further evidence from studies of pain to support the essential role of consciousness through the operation of different faculties. Instead of hypothesizing chains of unconscious inference, Cam proposes the interaction of a number of systems such as the cognitive-evaluative and motivational-affective, with differing computational resources. It is this kind of cooperative interaction among faculties which, Cam suggests, exhaustively accounts for 'higher cognition', though he concedes that this hypothesis is highly speculative and requires further support.

8. MICHAEL ARBIB: 'MODULARITY, SCHEMAS AND NEURONS: A CRITIQUE OF FODOR'

Arbib takes up this question of modularity and the general approach we have just seen, specifically challenging the account offered by Fodor,[25] and arguing, contrary to the previous article, that cognitive science needs a modular decomposition finer than that suggested by considerations purely at the level of such functional analysis. More particulary, Arbib points out that there is a continual tension between functionally determined decomposition into modules and structural decomposition since there is in general no one-to-one mapping between the analyses at these different levels. This distinction between functional and structural decomposition corresponds more or less to the top-down and bottom-up approaches respectively, and Arbib observes that ideally brain theory incorporates a cycle of both top-down and bottom-up modelling, in which there is the attempt to map functionally identified units onto brain regions.

Arbib begins by placing the problem in its historical context, citing the classical work of neurologists of the nineteenth century as the source of localization in terms of large interacting, functionally identified brain regions. Around the same time, the neurophysiological discoveries of Cajal and Sherrington established the neuron as the smallest functional component, and the problem was to discover some appropriate decomposition relating these two levels of analysis. Thus, Arbib cites more recent work which established the idea of the brain as an interconnected set of modules, intermediate between neurons and gross anatomical brain regions. Arbib points out that this recent and standard idea of decomposing a complex system into interacting subsystems is to be understood as *structural* insofar as the subsystems can be correlated with actual physical substructures of the brain. Noting that neuroscientists have typically referred to such *physical* substructures as 'modules', Arbib distinguishes this usage from Fodor's sense of the term which is intended to mean a *functionally* identified subsystem and conceived as meeting constraints beyond the usual ones.

Arbib offers detailed discussion of certain neuroscience research which serves to illustrate his critique of Fodor's modularity account. In particular, among Fodor's criteria for identifying modules, Arbib focuses upon the two claims that they are domain-specific and computationally autonomous, and he further challenges Fodor's taxonomy of the mind in terms of transducers, input systems and central processor. Looking at specific models of visual perception and visumotor coordination, Arbib demonstrates the manner in which functional modules can be identified. In particular, he shows that we are led to much finer-grained modular decomposition than suggested by Fodor. Whereas Fodor takes the visual system in its entirety as a single module, Arbib suggests postulating modules which subserve specific tasks as exemplified by depth perception or prey-recognition in the frog. At the same time, Arbib employs these researches to indicate the way in which other constraints proposed by Fodor are violated.

Fodor's account postulates a dramatic contrast between input systems and central processes, with only the former having a modular structure, whereas central systems are taken to lack such decomposability. As a consequence of this lack of modularity in central processes, Fodor suggests that it is unlikely that cognitive science will come to understand them. Since, in fact, Fodor believes that there is little direct evidence available bearing on the structure of central processes, his argument is based partly on the analogy of belief-fixation as a central process with scientific confirmation. It is the characteristic properties of such a scientific model which suggest the grounds for pessimism about central processes. Arbib challenges this analogy with scientific theorising and finally offers reasons for more modularity in central processes and, therefore, more optimistic conclusions about the ultimate prospects of cognitive science in this central domain.

9. HUW PRICE: 'ACTION EXPLANATION AND THE NATURE OF MIND'

As we have already seen, the developments in cognitive science have given rise to a debate concerning the status and ultimate fate of the large store of commonsense generalisations about human behaviour which have come to be labelled 'folk psychology'. On the one side there are those like Jerry Fodor who are staunch defenders of folk psychology and its propositional attitudes. On the other side, however, there are philosophers such as Stich and the Churchlands who reject folk psychology as stagnant, naive theory to be eliminated in a utopian cognitive science.

Against these critics Fodor has argued that there is little to compare with the predictive power of folk psychology which adverts to concepts such as belief and desire in the aetiology of behaviour. Fodor[26] has drawn attention to the remarkable predictive power of the commonsense generalisations of propositional attitude psychology and the fact that it works not just in highly constrained laboratory conditions, but among all the variability in everyday life. He observes: "If we could do that well with predicting the weather, no one would ever get his feet wet. Yet the aetiology of the weather is surely child's play compared with the causes of behaviour".[27] Furthermore, the importance of propositional attitude psychology within current cognitive science, in Fodor's view, is that it is precisely this kind of analysis which is vindicated by the computational approach. Fodor explains:

> (. . .) computational theories of mental processes don't replace the commonsense story about propositional attitudes and their behavioural effects. Rather, what a computational theory does is to make clear the mechanism of intentional causation; to show how it is (nomologically) possible that purely computational – indeed, purely physical – systems should act out of their beliefs and desires.[28]

It is in this context that Huw Price's paper is of value in taking up a detailed examination of certain problems concerning the role of beliefs and desires in the explanation of action. Beginning with David Hume's distinction between reason and passion, Price explores contemporary variants of Hume's idea that reason or belief is motivationally inert and cannot provide an exhaustive account of an action. Further appeal must be made to passions or desires in the causation of behaviour as Fodor insists and as reflected in his convenient metaphor of 'belief boxes' or 'intention boxes'. He explains that the difference between having a token mental symbol in one box or the other corresponds precisely to the difference in causal roles of beliefs and desires.[29]

The point of Price's critique is to challenge this two-box picture of cognitive architecture, particularly to the extent that it purports to be a quite general claim about arbitrary cognitive systems and not merely a parochial one about the contingent facts of human cognition. Additional interest derives from Price's approach to this problem, which is to explore the way in which essentially the same questions are raised in recent discussions of demonstrative and indexical beliefs. Accordingly, Price argues against the view that actions depend on demonstrative or indexical beliefs and then argues further that the contemporary claims for Hume's principle which rest on similar moves fail for similar reasons. At worst, Price suggests that indexicals and demonstratives reflect only the contingent limitations of our own particular systems, and even then not all the time; he suggests that one can easily imagine robots constructed with specific features which would obviate the need for indexical complexities. In considering the parallels with belief and desires, Price argues that analogous considerations eliminate reasons to accept Hume's distinction and to entertain the possibility in general of intrinsically rationalizing beliefs. That is, not just its empirical possibility, but the very cogency of Hume's distinction is challenged and Price concludes that, as cognitive scientists we lack a justification for the traditional two-box taxonomy of cognitive states.

CONCLUSION

The essays collected in this volume are representative of the main currents of thought among philosophers concerned with cognitive science. In particular, whatever its fate, the insights of folk psychology are no longer generally thought to exhaust the content and resources available to philosophy. To be sure, there is not a complete unanimity on this point but the general trend is clear enough as evidenced in the papers collected here and suggested by our title: that is, beyond the naive speculations founded on conceptual analysis, ordinary language or commonsense, philosophers today must have a detailed, technical understanding of 'computers, brains *and* minds'.

The last fifteen or twenty years which have seen the unfolding of the 'cognitive revolution' have supported T.S. Kuhn's[30] observation that scientific revolutions occur not because the practitioners of a discipline come to accept a new paradigm, but rather because a new generation grows up convinced of its truth. The interdisciplinary character of cognitive science reflects something of this insight as the specialists of one generation are gradually being overtaken by the converts to the multi-faceted cognitive science. In centres of cognitive science activity in the USA and elsewhere, there has already evolved a high degree of commonality among the specialists in different disciplines, to the extent that they can talk meaningfully about the ideas and theories of relevant fields other than their own. In this sense a true cognitive science 'creole' has developed from the patois and pidgin of earlier communication across academic boundaries. However, it must be admitted that in the Australian region serious dialogue among neurophysiologists, computer scientists, philosophers and others has not yet become a reality. Neuroscientists watching the spike trains from single cell recordings are skeptical of the value of computer science to their concerns; computer scientists in turn are doubtful about the possible relevance of psychologists or neuroscientists to their programs; and everyone is suspicious of philosophers. Nevertheless, the initiatives in cognitive science leading to the preparation of this collection have been significant steps in the right direction. For example, while no less skeptical, the computer scientists are at least yelling at the philosophers across the seminar room. This is undeniably progress, and beats ignoring us. If they are still wondering what philosophers are doing in all of this, as Jerry Fodor has quipped, "tell 'em we invented it!".

It is certain that the philosophical discussions in this volume will dispel the persistent stereotype of philosophers as woolly-headed, armchair theorists doubting that tables and chairs are real. Moreover, as we have observed earlier, we need only go back to Descartes to prove Fodor's point. So, in addition to the intrinsic value and interest of the volume, the editors hope that it will contribute to fostering a more cooperative conception among the practitioners of the various specialist disciplines, especially in the region of its origin. Not least of all, the volume will indicate clearly the importance of philosophers to this enterprise.

NOTES

[1] J.A. Fodor (1975) *The Language of Thought*, Crowell.
[2] *Ibid., p. 68.*
[3] *Ibid., p. 73.*
[4] J.A. Fodor (1981) 'Propositional Attitudes', *RePresentations*, Bradford/MIT, and J.A. Fodor and Z. Pylyshyn (1988) 'Connectionism and Cognitive Architecture', *Cognition* **XXVIII**, pp. 3–71.
[5] P.S. Churchland (1980) 'Language, Thought and Information Processing', *Nous* **XIV**, pp. 147–170.

[6] J.A. Fodor (1980) 'Methodological Solipsism Considered as a Research Strategy in Cognitive Psychology', *The Behavioural and Brain Sciences* **III**, pp. 63–109.
[7] J.A. Fodor (1987) *Psychosemantics*, Bradford/MIT.
[8] See *op. cit.* (note 1), p. 37.
[9] N. Chomsky (1980) *Rules and Representations*, Blackwell.
[10] See *op. cit.* (note 1), p. 27.
[11] J.A. Fodor (1983) *The Modularity of Mind*, Bradford/MIT.
[12] T. Burge (1986) 'Individualism and Psychology', *The Philosophical Review* **XCV**, No. 1, pp. 3–45.
[13] See *op. cit.* (note 1), p. 64.
[14] See *op. cit.* (note 5).
[15] D.E. Rumelhart and J.L. McLelland (eds.), (1986) *Parallel Distributed Processing* **I** and **II**, MIT.
[16] R.N. Shepard and J. Metzler (1971) 'Mental Rotation and Three Dimensional Objects', *Science* **CLXXI**, pp. 701–703.
[17] N. Block (ed.) (1981) *Imagery*, Bradford/MIT.
[18] S. Gaukroger (ed.) (1980) *Descartes: Philosophy, Mathematics and Physics*, Harvester Press.
[19] D.M. Clarke (1982) *Descartes' Philosophy of Science*, Manchester University Press.
[20] R. Descartes (1969) *Philosophical Writings*, E. Anscombe and P.T. Geach (eds.), Nelson, pp. 242–246.
[21] M. Brand (1986) *Intending and Acting*, Bradford/MIT.
[22] G.A. Miller, E. Galanter and K. Pribram (1960) *Plans and the Structure of Behaviour*, Holt, Rinehart and Winston.
[23] See *op. cit.* (note 11).
[24] H. Gardner (1983) *Frames of Mind: The Theory of Multiple Intelligences*, Basic Books.
[25] See *op. cit.* (note 11).
[26] J.A. Fodor (1986) 'Banish disContent' in J. Butterfield (ed.) *Language, Mind and Logic*, Cambridge University Press.
[27] *Ibid. p. 2.*
[28] *Ibid. p. 3.*
[29] See *op. cit.* (note 7).
[30] T.S. Kuhn (1962) *The Structure of Scientific Revolutions*, University of Chicago Press.

Centre for Cognitive Science
University of New South Wales, Sydney, Australia

J.A. FODOR

WHY THERE *STILL* HAS TO BE A LANGUAGE OF THOUGHT

"But why", Auntie asked with perceptible asperity, "does it have to be a *language*?" Auntie speaks with the voice of the Establishment, and her intransigence is something awful. She is, however, prepared to make certain concessions in the present case. First, she concedes that there are beliefs and desires and that there is a matter of fact about their intentional contents; there's a matter of fact, that is to say, about which proposition the intentional object of a belief or a desire is. Second, Auntie accepts the coherence of physicalism. It may be that believing and desiring will prove to be states of the brain, and if they do that's OK with Auntie. Third, she is prepared to concede that beliefs and desires have causal roles, and that overt behavior is typically the effect of complex interactions among these mental causes. (That Auntie was raised as a strict behaviorist goes without saying. But she hasn't been quite the same since the 60's. Come to think of it, which of us has?) In short, Auntie recognizes that psychological explanations need to postulate a network of causally related intentional states. "But why", she asks with perceptible asperity, "does it have to be a *language*?" Or, to put it more succinctly than Auntie often does, what – over and above mere Intentional Realism – does the Language of Thought Hypothesis buy? That is what this paper is about.[1]

A prior question: What – over and above mere Intentional Realism – does the Language of Thought Hypothesis *claim*? Here, I think, the situation is reasonably clear. To begin with, LOT wants to construe propositional attitude tokens as relations to symbol tokens. According to standard formulations, to believe that *P* is to bear a certain relation to a token of a symbol which means that *P*. (It is generally assumed that tokens of the symbols in question are neural objects and that the relation in question is 'computational'. But these assumptions won't be urgent in the present discussion.) Now, symbols have intentional contents and their tokens are physical objects in all the known cases. And – qua physical objects – symbol tokens are the right sorts of things to exhibit causal roles. So there doesn't seem to be anything that LOT wants to claim *so far* that Auntie needs to feel up-tight about. What, then, exactly is the issue?

Peter Slezak and W.R. Albury (eds.), Computers, Brains and Minds, 23–46

Here's a way to put it. Practically everybody thinks that the *objects* of intentional states are in some way complex: for example, that what you believe when you believe that John is late for dinner is something composite whose elements are – as it might be – the concept of John and the concept of being late for dinner (or – as it might be – John himself and the property of being late for dinner). And, similarly, what you believe when you believe that *P&Q* is also something composite, whose elements are – as it might be – the proposition that *P* and the proposition that *Q*.

But the (putative) complexity of the *intentional object* of a mental state does not, of course, entail the complexity of the mental state itself. It's here that LOT ventures beyond mere Intentional Realism, and it's here that Auntie proposes to get off the bus. LOT claims that *mental states* – and not just their propositional objects – *typically have constituent structure*. So far as I can see, this is the *only* real difference between LOT and the sorts of Intentional Realism that even Auntie admits to be respectable. So a defense of LOT has to be an argument that believing and desiring are typically structured states.

Consider a schematic formulation of LOT that's owing to Steven Schiffer. There is, in your head, a certain mechanism, an INTENTION BOX. To make the exposition easer, I'll assume that every intention is the intention to make some proposition true. So then here's how it goes in your head, according to this version of LOT, when you intend to make it true that *P*. What you do is, you put into the intention box a token of a mental symbol that MEANS that *P*. And what the box does is: it churns and gurgles and computes and causes and the outcome is that you behave in a way that (*ceteris paribus*) makes it true that *P*.

So, for example, suppose I intend to raise my left hand (I intend to make true the proposition that I raise my left hand). Then what I do is, I put in my intention box a token of mental symbol that means 'I raise my left hand'. And then, after suitable churning and gurgling and computing and causing, my left hand goes up. (Or it doesn't, in which case the *ceteris paribus* condition must somehow not have been satisfied.) Much the same story would go for my intending to become the next King of France, only in that case the gurgling and churning would continue appreciably longer.

Now, it's important to see that, although this is *going* to be a Language of Thought story, it's not a Language of Thought story yet. For, so far, all we have is what Intentional Realists qua Intentional

Realists (including Auntie qua Auntie) are prepared to admit: viz. that there are mental states that have associated intentional objects (like, for example, the state of having a symbol that means 'I raise my left hand' in my intention box) and that these mental states that have associated intentional objects also have causal roles (for example, my being in one of these causes my left hand to rise). What makes the story a Language of Thought story, and not just an Intentional Realist story, is the idea that these mental states that have content also have syntactic structure – constituent structure, as I'll sometimes say – that's appropriate to the content that they have. For example, it's compatible with the story I told above that what I put in the intention box when I intend to raise my left hand is a *rock*; so long as it's a rock that's semantically evaluable. Whereas, according to the LOT story, what I put in the intention box has to be something like a *sentence*; in the present case, it has to be a formula which contains, *inter alia*, an expression that denotes me and an expression that denotes my left hand.

Similarly, on the merely Intentional Realist story, what I put in the intention box when I intend to make it true that I raise my left hand and hop on my right foot might also be a rock (though not, of course, the same rock since the intention to raise one's left hand is not the same as the intention to raise one's left hand and hop on one's right foot). Whereas, according to the LOT story, if I intend to raise my left hand and hop on my right foot, I must put into the intention box a formula which contains, *inter alia*, a subexpression that means *I raise my left hand* and a subexpression that means *I hop on my right foot*.

It's important to see – indeed, it generates the issue that this paper is about – that Intentional Realism doesn't logically require the LOT story; it's no sort of *necessary* truth that only formulas – only things that have syntactic structure – are semantically evaluable. No doubt it's puzzling how a rock (or the state of having a rock in your intention box) could have propositional objects; but then, it's *no less* puzzling how a formula (or the state of having a formula in your intention box) could have a propositional object. It is, in fact, approximately equally puzzling how *anything* could have a propositional object, which is to say that it's puzzling how Intentional Realism could be true. For better or for worse, however, Auntie and I are both assuming that Intentional Realism *is* true. The question we're arguing about isn't, then, whether mental states have a semantics. Roughly, it's whether they have a syntax. Or, if you prefer, it's whether they have a *combinatorial* semantics; the kind of semantics in which there are (relatively) complex expressions whose content is determined, in some regular way, by the content of their (relatively) simple parts.

So here, to recapitulate, is what the argument is about: Everybody thinks that mental states have intentional objects; everybody thinks that the intentional objects of mental states are characteristically complex – in effect, that propositions have parts; everybody thinks that mental states have causal roles; and, for present purposes at least, everybody is a functionalist, which is to say that they all hold that mental states are individuated by reference to their causal powers. (This is, of course, implicit in the talk about 'intention boxes' and the like: To be – metaphorically speaking – in the state of having such and such a rock in your intention box is just to be – literally speaking – in a state that is the normal cause of certain sorts of effects and/or the normal effects of certain sorts of causes.) What's at issue, however, is the internal structure of these functionally individuated states. Auntie thinks they have none; only the *intentional objects* of mental states are complex. I think they constitute a language; roughly, the syntactic structure of mental states mirrors the semantic relations among their intentional objects. If it seems to you that this dispute among Intentional Realists is just a domestic squabble, I agree with you. But so was the Trojan War.

In fact, the significance of the issue comes out quite clearly when Auntie turns her hand to cognitive architecture. It is quite natural, given Auntie's philosophical views, for her to think of the mind as a sort of directed graph; the nodes correspond to semantically evaluable mental states, and the paths correspond to the causal connections between these states. To intend, for example, that *P&Q* is to be in a state that has a certain pattern of (dispositional) causal relations to the state of intending that *P* and to the state of intending that *Q*. (Eg. being in the first state is normally causally sufficient for being in the second and third.) We could diagram this relation in a familiar way, as shown in Figure 1.

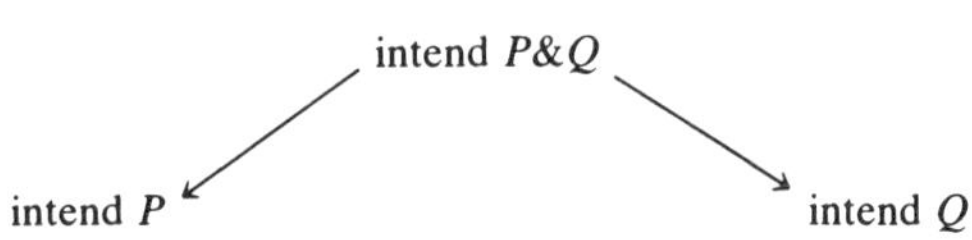

Fig. 1

It's of central importance to understand that, in this sort of architecture, the relation between – as it might be – intending that *P&Q* and intending that *P* is a matter of *connectivity* rather than *constituency*. You can see this instantly when you compare what's involved in intending that *P&Q* on the LOT story. On the LOT story, intending that *P&Q* requires having a sentence in your intention box – or, if you like, in a register or on a tape – one of whose parts is a token of the very same type that gets tokened in the intention box when you intend that *P*, and another of which is a token of the very same type that gets tokened in the intention box when you intend that *Q*. (Of course, on *either* story, the intention that *P&Q* is a normal cause of the intention that *P* and the intention that *Q*; and, qua functionalist, both stories take such causal connections to individuate the states that they hold among. In effect, the argument is over whether you need *both* connectivity and constituency relations to model the mind, or whether you can make do with connectivity alone.)

So, it turns out that the philosophical disagreement about whether there is a Language of Thought corresponds quite closely to the disagreement – current among cognitive scientists – about the appropriate architecture for mental models. Good Old Auntie is, of all things, a New Connectionist; and arguments that suggest that mental states have constituent structure *ipso facto* favor Turing/Von Neumann architecture as against the network picture.[2]

In what follows I propose to sketch three such arguments; three reasons for believing that cognitive states – and not just their intentional objects – typically have constituent structure. I don't suppose that these arguments are knock down; but I do think that, taken together, they ought to convince any Auntie who isn't *parti pris*.

First, however, I'd better 'fess up to a metaphysical prejudice that all three arguments assume. I don't believe that there are intentional mechanisms. That is: I don't believe that contents *per se* determine causal roles. In consequence, it's got to be possible to tell the whole story about mental causation (the whole story about the implementation of the generalizations that belief/desire psychologies articulate) *without referring to the intentional properties of the mental states that such generalizations subsume*. Suppose, in particular, that there is something about their causal roles that requires token mental states to be complex. Then I'm assuming that it does *not* suffice to satisfy this requirement that these mental states should have *complex intentional objects*.

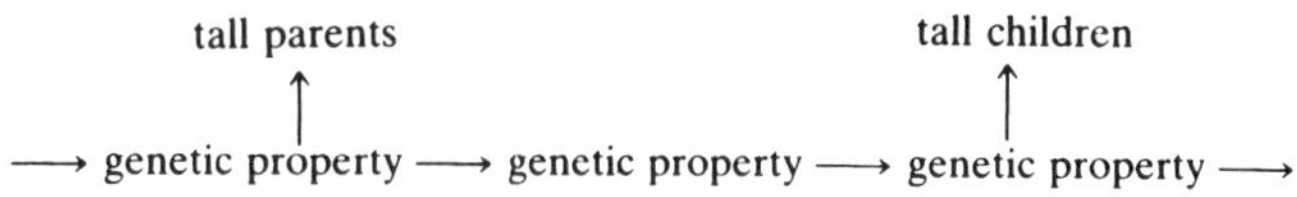

Fig. 2

This is not, by the way, any sort of epiphenomenalism; or, if it is, it's patently a harmless sort. There are plentry of cases in the respectable sciences where a law connects a pair of properties, but where the properties that the law connects *don't figure in the story about how the law is implemented*. So, for example, it's a law, more or less, that tall parents have tall children. And there's a pretty neat story about the mechanisms that implement that law; but the property of *being tall* doesn't figure in the story about the implementation; all that figures in that story is *genetic* properties. So you get something that looks like Figure 2, where the arrows indicate routes of causation.

The moral is that, even though it's true that psychological laws generally pick out the mental states that they apply to by specifying the intentional contents of the states, it *doesn't* follow that intentional properties figure in psychological mechanisms. And, while I'm prepared to sign on for counterfactual-supporting intentional generalizations, I balk at intentional causation. There are two reasons I can offer to sustain this prejudice (though I suspect that the prejudice goes deeper than the reasons). One of them is technical and the other is metaphysical.

Technical reason: If thoughts have their causal roles in virtue of their contents *per se*, then two thoughts with identical contents ought to be identical in their causal roles. And we know that this is wrong; we know that causal roles *slice things thinner* than contents do. The thought that $-P$, for example, has the same content as the thought that P on any notion of content that I can imagine defending; but the effects of entertaining these thoughts are nevertheless not guaranteed to be the same. Take a mental life in which the thought that $P \ \& \ (P \longrightarrow Q)$ immediately and spontaneously gives rise to the thought that Q; there is *no* guarantee that the thought that $-- \ \& \ P \ (P \rightarrow Q)$ immediately and spontaneously gives rise to the thought that Q in that mental life.

Metaphysical reason: It looks as though intentional properties essentially involve relations between mental states and *merely possible* contingencies. For example, it's plausible that for a thought to have the content THAT SNOW IS BLACK is for that thought to be related, in a certain way, to the possible (but non-actual) state of affairs in which snow is black; viz. it's for the thought to be true just in case that state of affairs obtains. Correspondingly, what distinguishes the content of the thought that snow is black from the content of the thought that grass is blue is the truth values that these thoughts have in possible but non-actual worlds.

Now, the following metaphysical principle strikes me as plausible: The causal powers of a thing are not affected by its relations to merely possible entities; only relations to *actual* entities affect causal powers. It is, for example, a determinant of my causal powers that I am standing on the brink of a high cliff. But it is *not* a determinant of my causal powers that I am standing on the brink of a possible-but-non-actual high cliff; I can't throw myself off one of *those*, however hard I try.[3]

Well, if this metaphysical principle is right, and if it's right that intentional properties essentially involve relations to nonactual objects, then it would follow that intentional properties are not per se determinants of causal powers, hence that there are no intentional mechanisms. I admit, however, that that is a fair number of ifs to hang an intuition on.

OK, now for the arguments that mental states, and not just their intentional objects, are structured entities.

1. A METHODOLOGICAL ARGUMENT

I don't, generally speaking, much like methodological arguments; who wants to win by a TKO? But, in the present case, it seems to me that Auntie is being a little unreasonable even by her own lights. Here is a plausible rule of nondemonstrative inference that I take her to be at risk of breaking:

Principle *P*: Suppose there is a kind of event $c1$ of which the normal effect is a kind of event $e1$; and a kind of event $c2$ of which the normal effect is a kind of event $e2$; and a kind of event $c3$ of which the normal effect is a complex event $e1$ & $e2$. Viz:

$$c1 \longrightarrow e1$$
$$c2 \longrightarrow e2$$
$$c3 \longrightarrow e1 \ \& \ e2$$

Then, *ceteris paribus*, it is reasonable to infer that $c3$ is a complex event whose constituents include $c1$ and $c2$.

So, for example, suppose there is a kind of event of which the normal effect is a bang and a kind of event of which the normal effect is a stink, and a kind of event of which the normal effect is that kind of a bang and that kind of a stink. Then, according to *P*, it is *ceteris paribus* reasonable to infer that the third kind of event consists (*inter alia*) of the co-occurence of events of the first two kinds.

You may think that this rule is arbitrary, but I think that it isn't; *P* is just a special case of a general principle which untendentiously requires us to prefer theories that *minimize accidents*. For, if the etiology of events that are *e1* and *e2* does not somehow include the etiology of events that are *e1* but not *e2*, then it must be that there are *two* ways of producing *e1* events; and the covergence of these (*ex hypothesi*) distinct etiologies upon events of type *e1* is, thus far, unexplained. (It won't do, of course, to reply that the convergence of two etiologies is only a very *little* accident. For, in principle, the embarrassment *iterates*. Thus, you can imagine a kind of event *c4*, of which the normal effect is a complex event *e1* & *e6* & *e7*; and a kind of event *c5*, of which the normal effect is a complex event *e1* & *e10* & *e12*. . . . etc. And now, if *P* is flouted, we'll have to tolerate a *four-way* accident. That is, barring *P* – and all else being equal – we'll have to allow that theories which postulate four kinds of causal histories for *e1* events are just as good as theories which postulate only one kind of causal history for *e1* events. It is, to put it mildly, hard to square this with the idea that we value our theories for the generalizations they articulate.)

Well, the moral seems clear enough. Let *c1* be intending to raise your left hand, and *e1* be raising your left hand; let *c2* be intending to hop on your right foot and *e2* be hopping on your right foot; let *c3* be intending to raise your left hand and hop on your right foot, and *e3* be raising your left hand and hopping on your right foot. Then the choices are: *either* we respect *P* and hold that events of the *c3* type are complexes which have events of type *c1* as constituents, *or* we flout *P* and posit two etiologies for *e1* events, the convergence of these etiologies being, thus far, accidental. I repeat that what's at issue here is the complexity of mental events and not the complexity of their intentional objects. *P* is a principle that constrains etiological inferences, and – according to the prejudice previously confessed to – the intentional properties of mental states are *ipso facto not* etiological.

But we're not home yet. There's a way out that Auntie has devised; she is, for all her faults, a devious old dear. Auntie could accept *P* but deny that (for example) raising your left hand counts as *the same sort of* event on occasions when you *just* raise your left hand as it does on occasions when you raise your left hand while you hop on your right foot. In effect, Auntie can avoid admitting that *intentions* have constituent structure if she's prepared to deny that *behavior* has constituent structure. A principle like *P*, which governs the assignment of etiologies to complex events, will be vacuously satisfied in psychology if no behaviors are going to count as complex.

But Auntie's back is to the wall; she is, for once, constrained by vulgar fact. Behavior does – very often – exhibit constituent structure, and that it does is vital to its explanation, at least as far as anybody knows. Verbal behavior is the paradigm, of course; everything in linguistics, from phonetics to semantics, depends on the fact that verbal forms are put together from recurrent elements; that, for example, [oon] occurs in both 'Moon' and 'June'. But it's not just verbal behavior for whose segmental analysis we have pretty conclusive evidence; indeed, it's not just *human* behavior. It turns out, for one example in a plethora, that bird song is a tidy system of recurrent phrases; we lose 'syntactic' generalizations of some elegance if we refuse to so describe it.

To put the point quite generally, psychologists have a use for the distinction between segmented behaviors and what they call 'synergisms'. (Synergisms are cases where what appears to be behavioral elements are in fact 'fused' to one another, so that the whole business functions as a unit; as when a well practiced pianist plays a fluent arpeggio.) Since it's empirically quite clear that not all behavior is synergistic, it follows that Auntie may not, in aid of her philosophical prejudices, simply help herself to the contrary assumption.

Now we *are* home. If, as a matter of fact, behavior is often segmented, then principle *P* requires us to prefer the theory that the causes of behavior are complex over the theory that they aren't, all else being equal. And all else *is* equal to the best of my knowledge. For, if Auntie has any *positive* evidence against the LOT story, she has been keeping it very quiet. Which wouldn't be at all like Auntie, I assure you.

ARGUMENT 2: PSYCHOLOGICAL PROCESSES (OR 'WHY AUNTIE CAN'T HAVE IT FOR FREE')

In the cognitive sciences mental symbols are the rage. Psycholinguists, in particular, often talk in ways that make Auntie simply livid. For, example, they say things like this: 'When you understand an utterance of a sentence, what you do is construct a *mental representation* [*sic*; emphasis mine] of the sentence that is being uttered. To a first approximation, such a representation is a parsing tree; and this parsing tree specifies the constituent structure of the sentence you're hearing, together with the categories to which its constituents belong. Parsing trees are constructed left to right, bottom to top, with restricted look ahead . . .' and so forth, depending on the details of the psycholinguist's story. Much the same sort of examples could be culled from the theory of vision or, indeed, from any other area of recent perceptual psychology.

Philosophical attention is hereby directed to the logical form of such theories. They certainly look to be quantifying over a specified class of mental objects: in the present case, over parsing trees. The usual apparatus of ontological commitment – existential quantifiers, bound variables and such – is abundantly in evidence. So you might think that Auntie would argue like this: 'When I was a girl ontology was thought to be an *a priori* science; but now I'm told that view is out of fashion. If, therefore, psychologists say that there are mental representations, then I suppose that there probably are. I therefore subscribe to the Language of Thought hypothesis'. That is not, however, the way that Auntie actually does argue. Far from it.

Instead, Auntie regards Cognitive Science in much the same light as Sodom, Gomorrah and Los Angeles. If there is one thing that Auntie believes in in her bones, it is the ontological promiscuity of psycholinguists. So, in the present case, although psycholinguists may *talk as though* they were professionally committed to mental representations, Auntie takes that to be *loose* talk. Strictly speaking, she explains, the offending passages can be translated out with no loss to the explanatory/predictive power of psychological theories. Thus, an ontologically profligate psycholinguist may speak of perceptual processes that construct a parsing tree; say one which represents a certain utterance as consisting of a noun phrase followed by a verb phrase, as in Figure 3.

But Auntie recognizes no such processes and quantifies over no such trees. What she admits instead are: (1) the utterance under perceptual analysis (the 'distal' utterance, as I'll henceforth call it); and (2) a mental process which eventuates in the distal utterance being *HEARD AS* consisting of a noun phrase followed by a verb phrase. Notice that this ontologically purified account, though it recognizes mental states

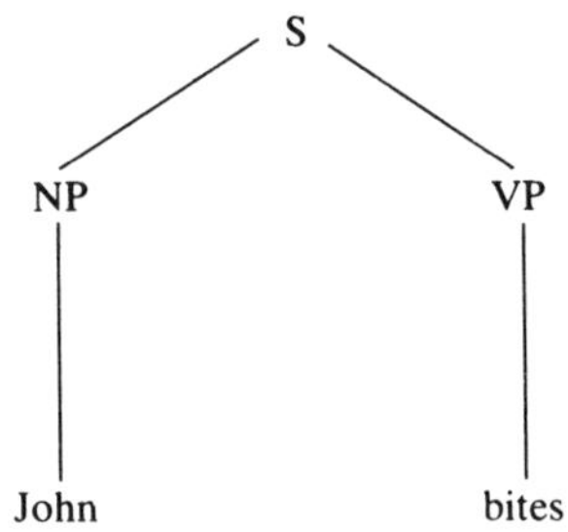

Fig. 3

with their intentional contents, does *not* recognize mental representations. Indeed, the point of the proposal is precisely to emphasize as live for Intentional Realists the option of postulating representational mental states and then crying halt. If the translations go through, then the facts which psychologists take to argue for mental representations don't actually do so; and if those facts don't, then maybe nothing does.

Well, but *do* the translations go through? On my view, the answer is that some do and others don't, and that the ones that don't make the case for a Language of Thought. This will take some sorting out.

Mental representations do two jobs in theories that employ them. First, they provide a canonical notation for specifying the intentional contents of mental states. But second, mental symbols constitute domains over which *mental processes* are defined. If you think of a mental process – extensionally, as it were – as a sequence of mental states each specified with reference to its intentional content, then mental representations provide a mechanism for the construction of these sequences; they allow you to get, in a mechanical way, from one such state to the next *by performing operations on the representations*.

Suppose, for example, that this is how it goes with English *wh*-questions: Such sentences have two constituent structures, one in which the questioned phrase is in the OBJECT position as per Figure 4, and one in which the questioned phrase is in the SUBJECT position, as per Figure 5.

And suppose that the psycholinguistic story is that the perceptual analysis of utterance of such sentences requires the assignment of these constituent structures in, as it might be, reverse order. Well, Auntie can tell *that* story *without* postulating mental representations; *a fortiori*

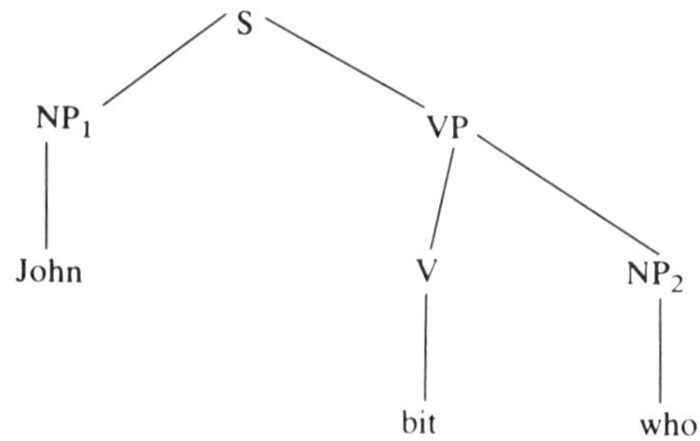

Fig. 4

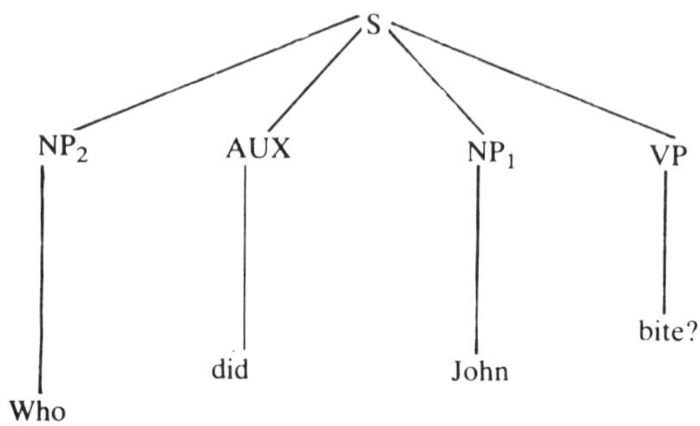

Fig. 5

without postulating mental representations that have constituent structure. She does so by talking about THE INTENTIONAL CONTENTS OF THE HEARER'S MENTAL STATES rather than the mental representations he constructs. "The hearer", Auntie says, "starts out by repsesenting the distal utterance as having 'John' in the subject position and a questioned NP in the object position; and he ends up representing the distal utterance as having these NPs in the reverse configuration. Thus we see that when it's *properly* constructed, claims about 'perceiving as' is all that talk about mental representation ever really comes to." Says Auntie.

But in saying this, it seems to me that Auntie goes too fast. For, what *doesn't* paraphrase out this way is the idea that the hearer gets from one of these representational states to the other BY MOVING A PIECE OF THE PARSING TREE (e.g. by moving the piece that represents 'who' as a constituent of the type NP_2). This untranslated part of the story isn't, notice, about what intentional contents the hearer entertains or the order in which he entertains them. Rather, it's about the mechanisms that mediate the transitions among his intentional states. Roughly, the story says that the mechanism of mental state transition is *computational*; and if the story's true, then (a) there must *be* parsing trees for mental processes to operate upon; and (b) these parsing trees need to have a kind of structure that will sustain talk of moving part of a tree while leaving the rest of it alone; in effect, they need to have constituent structure.

There's an aspect of the argument just presented that I particularly want to stress: The proposed inference deduces a language of thought from a computational story about how we understand *wh*-questions. It depends on assuming that understanding these questions requires the explicit mental representation of the constituent structure of sentences. But it does *not* assume the explicit representation of the rule of *wh*-movement. I emphasize this point in light of recent philosophical discussions of the role of mental representation in computational theories of mind; these have centered, almost exclusively, on issues about the representation of *rules and programs*. They have thereby occasioned much confusion.

Dennett, for example, makes the following point (see Dennett, (1978)[4]): You can have a chess playing machine that is (let's assume) correctly described as preferring to get the queen out early, but whose program contains no sentence (symbol, rule, expression, whatever) that means: *get the queen out early*. (In the case that Dennett has in mind, the fact that the machine does tend to get the queen out early is a sort of emergent out of the implementation of many far more detailed principles of play.) So we might say – and philosophers do say – that though the machine's behavior accords with the queen-early rule, the machine is nevertheless not *following* that rule. *A fortiori*, the behavioral facts about the machine give no support to the claim that the queen-early rule is internally represented. (Similarly, *mutatis mutandis*, the planets do not *follow* Kepler's laws as they proceed upon their rotations, and explaining their motions does not require an astronomical analogue of LOT).

This sort of point can be pushed further (and has been in the context of philosophical discussions of the linguistic evidence for the internal representation of grammars; see especially Stabler (1983); Matthews (1984);[5] Demopoulos and Matthews (1983)).[6] Here, very briefly, is how the argument goes: Suppose there is a machine that computes a function by making use of an explicit algorithm. Then there is always an equivalent 'hard-wired' machine in which the function computed is *not* explicitly represented. So even if you assume that the behavioral facts about an intentional system justify the claim that it is computing function *F*, still there is no inference to the explicit representation of *F* in the machine. Surprisingly, then, it looks as though the theory that the mind is a computer is actually *neutral* on the question whether there is a language of thought.

I am not, however, much moved by this argument; and the reason why I'm not should now be clear. A parsing machine needn't represent the rule of *wh*-movement explicitly; but if it's to run at all it must explicitly represent the structure of the sentences to which that rule applies. It can't compute without representing these structures because representations of structures are what it computes *on*. So if the mind is (*inter alia*) a parsing machine, then there must be explicit mental representation and there must be a language of thought. Or again, a chess playing computer need not explicitly represent the rules of chess, or even the tactical principles that govern its moves. But it *must* explicitly represent the state of the board at each stage of play; representations of states of the board are what its computations are *defined over*. So, if the mind is (*inter alia*) a chess playing machine, then again there must be explicit representation and again there must be a language of thought.

The moral: *programs* needn't be explicitly represented, but *computational domains* (roughly, data structures) have got to be. So, it's computational domains, not programs, that make the argument for a language of thought.

Here's another way to make this point. Of the many mental states we may have reason to attribute to an intentional system, some – but typically not all – will figure as links in causal chains of mental events. A rule of chess strategy (like the queen-early rule) or of grammar (like the move-*wh* rule) may play this sort of role; it will do so when the system computes *by consulting the rule*. But it won't do so in Dennett's sort of machine because, by hypothesis, the thought 'get your queen out early' *never occurs to Dennett's machine*; (similarly, Kepler's Laws never occur to the planets). *A fortiori*, occurrences of that thought are never links in the causal chains that generate the machine's moves, so an etiological account of the moves is not committed to the explicit representation of the rule. Conversely, however, if a computational psychology says that tokens of a mental state *are* links in causal chains, then *the intentional object of the state must be explicitly represented or the theory is false*. The operative principle is: 'No intentional causation without explicit representation'. And to *this* principle Dennett's case is *not* a counterexample.

This picture of how the computational story about mental processes connects with the Realist story about mental representation has a philosophically interesting consequence. Since the propositional object of a mental state must be explicitly represented whenever tokens of the state are causally involved, it follows that a computational theory must distinguish among *extensionally equivalent* states in all such cases. In the sort of example Dennett has in mind, there often isn't a matter of fact about just what intentional content the machine's rule has. Is it, for example, 'get the queen out early' or is it 'get the highest valued piece out early'? The two versions are extensionally equivalent, and since by hypothesis the rule is not internally represented, there would seem to be nothing to choose between these formulations. But you can imagine building a different sort of machine – one where the principles of play *are* consulted in the course of the computations – and here it *would* matter which rule the machine follows: Though the moves come out the same either way, *the course of the computations* would be different depending on how the rule is formulated. Causal involvement sets a lower bound on the explicitness of representation; and, in so doing, it also determines what counts as equivalence in attributions of intentional content.[7]

Well, to return to Auntie, I must now report a quirk of hers that I do not fully understand: She refuses to take seriously the ontological commitments of computational theories of mental processes. This is all the more puzzling because Auntie is usually content to play by the following rule: Given a well-evidenced empirical theory, either you endorse the entities that it's committed to, or you find a paraphrase that preserves the theory while dispensing with the commitments. Auntie holds that this is simply good deportment for a philosopher; and I, for once, agree with her completely. So, as we've seen, Auntie has a proposal for deontologizing the computational story about which state understanding a sentence is: She proposes to translate talk about trees in the head into talk about hearing utterances under descriptions, and that seems to be all right as far as it goes. But it doesn't go far enough because the ontological commitments of psychological theories are inherited not just from their account of mental states, cut also from their account of mental processes; and the computational account of mental processes would appear to be *ineliminably* committed to mental representations construed as structured objects.

The moral, I suppose, is that if Auntie won't bite the bullet, she will have to pay the piper. As things stand now, the cost of not having a Language of Thought is not having a theory of thinking. It's a striking fact about the philosophy of mind that we've indulged for the last fifty years or so that it's been quite content to pony up this price. Thus, while an 18th Century Empiricist – like Hume, say – took it for granted that a theory of cognitive *processes* (specifically, Associationism) would have to be the cornerstone of psychology, modern philosophers – like Wittgenstein and Ryle and Auntie – *have* no theory of thought to speak of. I do think this is appalling; how can you seriously hope for a good account of belief if you have no account of belief *fixation*? But I don't think it's entirely surprising. Modern philosophers who haven't been overt behaviorists have quite generally been covert behaviorists. And while a behaviorist can recognize mental states – which he identifies with behavioral dispositions – he has literally no use for cognitive processes like causal trains of thought. The last thing a behaviorist wants is mental causes ontologically distinct from their behavioral effects.

It may be that Auntie has not quite outgrown the behaviorist legacy of her early training. Anyhow, if you ask her what she's prepared to recognize in place of computational mental processes, she unblushingly replies (I quote): "Unknown Neural Mechanisms". (I think she may have gotten that from John Searle, whose theory of thinking it closly resembles.) If you then ask her whether it's not sort of unreasonable to prefer no psychology of thought to a computational psychology of thought, she affects a glacial silence. Ah well, there's nothing to be done with Auntie when she stands upon her dignity and strikes an Anglo-Saxon attitude; except to try a different line of argument.

ARGUMENT 3: PRODUCTIVITY AND SYSTEMATICITY

The classical argument that mental states are complex adverts to the productivity of the attitudes. There is a (potentially) infinite set of – for example – belief state types, each with its distinctive intentional object and its distinctive causal role. This is immediately explicable on the assumption that belief states have combinatorial structure; that they are somehow built up out of elements and that the intentional object and causal role of each such state depends on what elements it contains and how they are put together. The LOT story is, of course, a paradigm of this sort of explanation since it takes believing to involve a relation to a syntactically structused object for which a compositional semantics is assumed.

There is, however, a notorious problem with productivity arguments. The facts of mortality being what they are, not more than a finite part of any mental capacity ever actually gets exploited. So it requires idealization to secure the crucial premise that mental capacities really are productive. It is, for example, quite possible to deny the *productivity* of thought even while admitting that people are forever thinking new things. You can imagine a story – vaguely Gibsonian in spirit – according to which cognitive capacity involves a sort of 'tuning' of the brain. What happens, on this view, is that you have whatever experiences engender such capacities, and the experiences have Unknown Neurological Effects (these Unknown Neurological Effects being mediated, it goes without saying, by the corresponding Unknown Neurological Mechanisms) and the upshot is that you come to have a very large – but finite – number of, as it were, *independent* mental dispositions. E.g. the disposition to think that the cat is on the mat on some occasions; and the disposition to think that 3 is prime on other occasions; and the disposition to think that secondary qualities are epiphenomenal on other occasions . . . and so forth. New occasions might thus provoke novel thoughts; and yet the capacity to think wouldn't have to be productive. In principle it could turn out, after a lot of thinking, that your experience catches up with your cognitive capacities so that you actually succeed in thinking everything that you are able to. It's no good saying that you take this consequence to be absurd; Auntie doesn't.

In short, it needs productivity to establish that thoughts have combinatorial structure, and it needs idealization to establish productivity; so it's open to Somebody who doesn't want to admit productivity (because, for example, She doesn't like LOT) simply to refuse to idealize. This is, no doubt, an empirical issue in the very long run. Scientific idealization is demonstrably appropriate if it eventually leads to theories that are independently well confirmed. But vindication in the very long run is a species of cold comfort; perhaps there's a way to get the goodness out of productivity arguments *without* relying on idealizations that are plausibly viewed as tendentious. Here's how I propose to argue:

(a) There's a certain property that linguistic capacities have in virtue of the fact that natural languages have a combinatorial semantics.
(b) Thought has this property too.
(c) So thought too must have a combinatorial semantics.

Auntie remarks that this has the form of affirmation of the consequent. So be it; one man's affirmation of the consequent is another man's inference to the best explanation.

The property of linguistic capacities that I have in mind is one that inheres in the ability to understand and produce sentences. That ability is – as I shall say – *systematic*: By which I mean that *the ability to produce/understand some of the sentences is INTRINSICALLY connected to the ability to produce/understand many of the others*. You see the force of this if you compare learning a language the way we really do learn them with learning a language by memorizing an enormous phrase book. The present point isn't that phrase books are finite and can therefore exhaustively describe only *non*productive languages; rather it's that you can learn *any part* of a phrase book *without learning the rest*. Hence, on the phrase book model, it would be perfectly possible to learn that uttering the form of words 'Granny's cat is on Uncle Arthur's mat' is the way to say that Granny's cat is on Uncle Arthur's mat, and yet have *no idea* how to say that it's raining (or, for that matter, how to say that Uncle Arthur's cat is on Granny's mat). I pause to rub this point in. I know – to a first approximation – how to say *Who does his mother love very much?* in Korean; viz: 'ki-iy emma-ka nuku-lil mewu saranna-ci?' But since I did get this from a phrase book, it helps me not at all with saying anything else in Korean. In fact, I don't know how to say anything else in Korean; I have just shot my bolt.

Perhaps it's self evident that the phrase book story must be wrong about language acquisition because a speaker's knowledge of his native language is never like that. You don't, for example, find native speakers who know how to say in English that John loves Mary but don't know how to say in English that Mary loves John. If you did find someone in such a fix, you'd take that as presumptive evidence he's not a native English speaker but some sort of a tourist. (This is one important reason why it is so misleading to speak of the block/slab game that Wittengenstein describes in Paragraph 2 of the *Investigations* as a "complete primitive language"; to think of languages that way is precisely to miss the systematicity of linguistic capacities (to say nothing of their productivity).)

Notice, by the way, that systematicity (again like productivity) is a property of sentences but not of words. The phrase book model really *does* fit what it's like to learn the *vocabulary* of English since when you learn English vocabulary you acquire a lot of basically *independent* dispositions. So you might perfectly well learn that using the form of words 'cat' is the way to refer to cats and yet have no idea that using the form of words 'deciduous conifer' is the way to refer to deciduous conifers. My linguist friends tell me that there are languages – unlike English – in which the lexicon, as well as the syntax, is productive. It's candy from babies to predict that a native speaker's mastery of the vocabulary of such a language is always systematic. Productivity and systematicity run together; if you postulate mechanisms adequate to account for the one, you get the other automatically.

What sort of mechanisms? Well, the alternative to the phrase book story about acquisition depends on the idea, more or less standard in the field since Frege, that the sentences of a natural language have a combinatorial semantics (and, *mutatis mutandis*, that the lexicon does in languages where the lexicon is productive). On this view, learning a language is learning a perfectly general procedure for determining the meaning of a sentence from a specification of its syntactic structure together with the meanings of its lexical elements. Linguistic capacities *can't help but* be systematic on this account because, give or take a bit, the very same combinatorial mechanisms that determine the meaning of any of the sentences determine the meaning of all of the rest.

Notice two things: First, you can make these points about the systematicity of language without idealizing to astronomical computational capacities. *Productivity* is involved with our ability to understand sentences that are a billion trillion zillion words long. But *systematicity* involves facts that are much nearer home: such facts as the one we mentioned above, that no native speaker comes to be able to understand the form of words 'John loves Mary' except as he *also* comes to be able to understand the form of words 'Mary loves John'. Insofar as there are 'theory neutral' data to constrain our speculations about language, this surely ought to count as one of them.

Second, if the systematicity of linguistic capacities turns on sentences having a combinatorial semantics, the fact that sentences have a combinatorial semantics turns on their having constituent structure. You can't construct the meaning of an object out of the meanings of its constituents unless it *has* constituents. The sentences of English wouldn't have a combinatorial semantics if they weren't made out of recurrent words and phrases.

OK, so here's the argument: linguistic capacities are systematic, and that's because sentences have constituent structure. But cognitive capacities are systematic too, and that must be because *thoughts* have constituent structure. But if thoughts have constituent structure, then LOT is true. So I win and Auntie looses. Goody!

I take it that what needs defending here is the idea that cognitive capacities are systematic, *not* the idea that the systematicity of cognitive capacities implies the combinatorial structure of thoughts. I get the second claim for free for want of an alternative theory. So then, how do we know that cognitive capacities are systematic?

A fast argument is that cognitive capacities must be *at least* as systematic as linguistic capacities since the function of language is to express thought. To understand a sentence is to grasp the thought that its utterance standardly conveys; so it wouldn't be possible that everyone who understands the sentence 'John loves Mary' also understands the sentence 'Mary loves John' if it weren't that everyone who can *think the thought* that John loves Mary can also think the thought that Mary loves John. You can't have it that language expresses thought *and* that language is systematic unless you also have it that thought is as systematic as language is.

And that is quite sufficiently systematic to embarrass Auntie. For, of course, the systematicity of thought does *not* follow from what Auntie is prepared to concede: viz. from mere Intentional Realism. If having the thought that John loves Mary is just being in one Unknown But Semantically Evaluable Neural Condition, and having the thought that Mary loves John is just being in another Unknown But Semantically Evaluable Neural Condition, then it is – to put it mildly – not obvious why God couldn't have made a creature that's capable of being in one of these Semantically Evaluable Neural conditions but not in the other, hence a creature that's capable of thinking one of these thoughts but not the other. But if God could have made such a creature, then mere Intentional Realism doesn't explain the systematicity of thought; as we've seen, mere Intentional Realism is exhausted by the claim that there *are* Semantically Evaluable Neural Conditions.

To put it in a nutshell, what you need to explain the systematicity of thought appears to be Intentional Realism *plus LOT*. LOT says that having a thought is being related to a structured array of representations; and, presumably, to have the thought that John loves Mary is *ipso facto* to have access to the same representations, and the same representational structures, that you need to have the thought that Mary loves John. So, *of course* anybody who is in a position to have one of these thoughts is *ipso facto* in a position to have the other. LOT explains the systematicity of thought; mere Intentional Realism doesn't. Thus I refute Auntie!

Four remarks to tidy up: First, this argument takes it for granted that systematicity is *at least sometimes* a contingent feature of thought; that there are *at least some cases* in which it is logically possible for a creature to be able to entertain one but not the other of two content-related propositions.

I want to remain neutral, however, on the question whether systematicity is *always* a contingent feature of thought. For example, a philosopher who is committed to a strong 'inferential role' theory of the individuation of the logical concepts might hold that you can't, in principle, think the thought that (*P* or *Q*) unless you are able to think the thought that *P*. (The argument might be that the ability to infer (*P* or *Q*) from P is *constitutive of having* the concept of disjunction.) If this claim is right, then – to that extent – you don't need LOT to explain the systematicity of thoughts which contain the concept *or*; it simply *follows from* the fact that you can think that *P* or *Q* that you can also think that *P*.

Auntie is, of course, at liberty to try to explain *all* the facts about the systematicity of thought in this sort of way. I wish her joy of it. It seems to me perfectly clear that there could be creatures whose mental capacities constitute a proper subset of our own; creatures whose mental lives – viewed from our perspective – appear to contain gaps. If inferential role semantics denies this, then so much the worse for inferential role semantics.

Second: It is, as always, essential not to confuse the properties of the attitudes with the properties of their objects. I suppose that it *is* necessarily true that the *propositions* are 'systematic'; i.e. that if there is the proposition that John loves Mary, then there is also the proposition that Mary loves John. But that necessity is no use to Auntie since it doesn't explain the systematicity of our capacity to *grasp* the propositions. What LOT explains – and, to repeat, mere Intentional Realism does not – is a piece of our empirical psychology: the *de facto*, contingent connection between our ability to think one thought and our ability to think another.

Third: Many of Auntie's best friends hold that here is something *very* special about language; that it is *only* when we come to explaining linguistic capacities that we need the theoretical apparatus that LOT provides. But, in fact, we can kick the ladder away: we don't need the systematicity of language to argue for the systematicity of thought. All we need is that it is on the one hand true, and on the other hand not a *necessary* truth, that whoever is able to think that John loves Mary is *ipso facto* able to think that Mary loves John.

Auntie has the option of arguing the *empirical* hypothesis that thought is systematic only for creatures who speak a language. But think what it would mean for that hypothesis to be true. It would have to be quite usual to find animals that are capable of learning to respond selectively to situations such that *a R b*, but quite unable to learn to respond selectively to a situation such that *b R a* (so that you can teach the beast to select the picture with the triangle larger than the square, but you can't for the life of you teach it to select the picture with the square larger than the triangle.) I am not into rats and pigeons. But I once took a Comp. Psych. course, and I'm perfectly certain that animal minds aren't, in general, like that.

Fourth: Just a little systematicity of thought will do to make things hard for Auntie since, as previously remarked, mere Intentional Realism is compatible with there being no systematicity of thought at all. And this is just as well because, although we can be sure that thought is somewhat systematic we can't, perhaps, be sure of *just how* systematic it is. The point is that if we are unable to think the thought that *P*, then I suppose we must also be unable to think the thought that we are unable to think the thought that *P*. So it's at least arguable that, to the extent that our cognitive capacities are *not* systematic, the fact that they are not is bound to escape our attention. No doubt this opens up some rather spooky epistemic possibilities; but, as I say, it doesn't matter for the polemical purposes at hand. The fact that there are *any* contingent connections between our capacities for entertaining propositions is remarkable when rightly considered. I know of no account of this fact that isn't tantamount to LOT. And neither does Auntie.

So we've found at least three reasons for preferring LOT to mere Intentional Realism, and three reasons ought to be enough for anybody's Auntie. But is there any *general* moral to discern? Maybe there's this one: If you look at the mind from what has recently become the philosopher's favorite point of view, it's the semantic evaluability of mental states that looms large. What's puzzling about the mind is that anything *physical* could have satisfaction conditions, and the polemics that center around Intentional Realism are the ones that this puzzle generates. On the other hand, if you look at the mind from the cognitive psychologist's viewpoint, the main problems are the ones about mental processes. What puzzles psychologists is belief fixation and, more generally, the contingent, causal relations that hold among states of mind. The characteristic doctrines of modern cognitive psychology (including, notably, the idea that mental processes are computational) are thus largely motivated by problems about mental causation. Not surprisingly, given this divergence of main concerns, it looks to philosophers as though the computational theory of mind is mostly responsive to technical worries about mechanism and implementation; and it looks to psychologists as though Intentional Realism is mostly responsive to metaphysical and ontological worries about the place of content in the natural order. So, deep down, what philosophers and psychologists really want to say to one another is: 'Why do you care so much about *that*?'

Now as Uncle Hegel used to enjoy pointing out, the trouble with perspectives is that they are, by definition, *partial* points of view; the real problems are appreciated only when, in the course of the development of the World Spirit, the limits of perspective come to be transcended. Or, to put it another way, it helps to be able to see the whole elephant. In the present case, I think the whole elephant looks like this: The key to the nature of cognition is that mental processes preserve semantic properties of mental states; trains of thought, for example, are generally truth preserving, so if you start your thinking with true assumptions you will generally arrive at conclusions that are also true. The central problem about the cognitive mind is to understand how this is so. And my point is that *neither* the metaphysical concerns that motivate Intentional Realists, nor the problems about implementation that motivate cognitive psychologists suffice to frame this issue. To see this issue, you have to look at the problems about content and the problems about process *at the same time*. Thus far has the World Spirit progressed.

If Auntie's said it once, she's said it a hundred times: Children should play nicely together and respect each other's points of view. I do think Auntie's right about that.[8]

NOTES

[1] Auntie's not the only one who'd like to know; much the same question has been raised by Noam Chomsky, John Searle, Brian Loar, David Israel, Jon Barwise and John Perry, and Tyler Burge to name just a few. Auntie and I are grateful to all of the above for conversations which led to the present reflections. Also to Ned Block for characteristically perceptive comments on an earlier draft.

[2] It's worth emphasizing that the issue is not directly about Associationism. Classical Associationists (Hume, say) held that mental representations have constituent structure and, I suppose, a combinatorial semantics. (The image of a house *ipso facto* contains, as proper parts, images of proper parts of houses.) Classical Associationism is therefore on my side of the argument with Auntie and with New Connectionists.

[3] Notice – by contrast – that relations to nonactual entities can perfectly well be *constitutive of* causal powers: The solubility of this salt consists in such facts as that if there *were* water here, the salt would have dissolve in it. The point in the text, then, is that though relations to nonactual objects can figure in the analysis of a causal power, they can't be among its causal determinants. Nothing – causal powers included – can be an effect of a merely possible cause. (I'm grateful to Georges Rey for helping me get this sorted out.)

[4] D.C. Dennett (1978) 'A Cure for the Common Code?', *Brainstorms*, Bradford Books.

[5] R.J. Matthews (1984) 'Troubles with Representationalism', *Social Research* **LI**, No. 4, pp. 1065–1197. E.P. Stabler (1983) 'How Are Grammars Represented?', *The Behavioural and Brain Sciences* **VI**, pp. 391–421.

[6] W. Demopoulos and R.J. Matthews (1980) 'On the Hypothesis that Grammars are Mentally Represented', *The Behavioural and Brain Sciences* **III**, pp. 405–406.

[7] This is not, of course, a free lunch. You get a principle of individuation for content attribution but only if you assume a principle of individuation for courses of computation.

[8] A slightly different version of this paper appears as an appendix in J.A. Fodor (1987) *Psychosemantics*, Bradford/MIT.

Department of Philosophy
The Graduate Centre, City University of New York, U.S.A.

D.H. MELLOR

HOW MUCH OF THE MIND IS A COMPUTER?*

1. How much of the mind is a computer? Computational psychologists, who 'see psychology as the study of the various computational processes whereby mental representations are constructed, organised and transformed',[1] say that most if not all of it is. I think they are wrong. Most of the mind is not a computer: most mental processes are not computations.

Before arguing for this conclusion I must distinguish my question from others that may be confused with it, which I will not discuss. One is its converse: how much of a mind do computers have? So far not much; but for different reasons. Maybe in future they will have more; but I have nothing to say about that. Nor about the extent to which psychology should be computable, so that psychologists can use computer models of the mind. That question is not about the mind at all, but about theories of it. A computable theory has no doubt the virtue of being easier to study. But that virtue is not peculiar to theories of the mind; and it doesn't make theories that have it theories of computers: computer models of the mind no more imply that the mind is a computer than computer models of the economy imply that the economy is.

Setting these red herrings aside, what an answer to *my* question tells us about the mind depends on what computers are. On some conceptions of them, it tells us almost nothing. So to give the question a useful sense, I must first distinguish and discard those conceptions. And this will take some time, because it takes time to shake the conviction that we know what computers are because we are familiar with the general purpose programmable electronic digital computers that are their modern paradigms. They are what give our question its interest and its real sense: 'How much of the mind is to be explained in terms of such machines?'. And this sense seems clear because we have a largely agreed vocabulary for describing the working and uses of these machines: such terms as 'computation', 'representation', 'information', 'data', 'processing', 'program', 'syntax', 'semantics' and 'algorithm'. This all generates a spurious consensus about what computers are – spurious because different people read key terms in this list, notably 'information' and 'representation', very differently. Those differences in turn affect the meanings of other terms and hence the sense of our

Peter Slezak and W.R. Albury (eds.), Computers, Brains and Minds, 47–69

question. But before showing how that happens, and identifying the sense the question needs, I must sketch some agreed relations between the relevant terms.

2. First, all parties agree that a computer is defined by what it does, namely compute. It doesn't matter how it computes. When, for instance, one computation needs the results of others, it is immaterial whether the others are done in series (one after the other) or in parallel (together). Nor does it matter whether a computer computes with silicon chips or with brain cells. It need not even compute with matter; a spiritual computer is not a contradiction in terms. Whether mental processes are computations is independent of whether they are material. (So fortunately we can set that traditionally vexed question aside.)

Next, however computing is done, it is agreed to be the processing of information, as opposed to the processing of matter. A food processor making paté from its ingredients is processing matter: a word processor making a description of the paté from a description of its ingredients is processing information. What is the difference?

The main difference is that information is true or false, whereas matter just is. Pieces of information are what I shall call 'propositions', whether they are expressed or embodied in sentences, pictures, computers, beliefs or any other way. A proposition, e.g. *the earth is round*, corresponds to a state of affairs (the earth's being round), which may or may not obtain. If it does (if the earth *is* round), it is a fact, which makes the proposition true; if not, the proposition is false.

To say this is not to assume specific theories of propositions, states of affairs, facts or truth: e.g. a correspondence theory of truth. The only assumptions I make here are those I shall make explicitly. The main one is that propositions exist, or at least that we have attitudes (belief, desire, hope, fear, etc.) of which they are the *prima facie* contents: e.g. my belief that the earth is round. The assumption has been disputed, e.g. by Stich,[2] but like Fodor[3] and others, I stand by it. I cannot argue for it here, except to say that for the present we clearly need such propositional attitudes to make sense of the mind. Maybe psychology will eventually displace them, but I doubt it. (I doubt especially that computational psychology will; since, as we shall see, computation gets its own sense from the concept of belief.)

Another assumption I make is that besides being true or false, the contents of our beliefs also differ from matter by being abstract, with no location in space and time, and no causal powers. The earth must be somewhere, and facts about it will have causes and effects. But propositions about it, whatever they are, are nowhere in particular, and neither affect nor are affected by anything. Yet computers process propositions causally, in definite places and at definite times. They must therefore process them indirectly, by processing *tokens* of them that do have causes, effects and spatio-temporal location. A proposition therefore is a *type*: again unlike a piece of matter, the very same piece of information may be processed many times, and in any number of places at once. The information output by a single computation will therefore not be a new proposition, but a new token of a proposition, generated by a causal process from input tokens of other propositions.

This causal processing of information we may take to be deterministic. (Exceptions, e.g. outputting a random choice among many solutions to an equation, raise no special issues and only complicate the discussion.) There may be 'noise' in a deterministic computation, but there shouldn't be. The information output by it should be fixed by the information input. That is, it should be a *function* of it in the sense in which (e.g.) birthdays are a function of people but not *vice versa*: people – the birthday function's 'arguments' – have only one birthday each – its 'values' for those arguments; whereas many people share the same birthday. (This sense of 'function', since it figures largely in what follows, I must say at once has almost no connection with the concept of a function in biology, anthropology or the philosophy of mind.)

But a computation's 'information function' cannot be given just by its causal processing. Causal processes work only on intrinsic properties of the tokens involved. The properties may be chemical, or electrical, or even mental – e.g. being some kind of pain or visual sensation. The range of intrinsic properties is disputed: whether for instance it includes relational properties, like being hotter than something. But no one thinks it includes being a token of anything like a proposition, say that the earth is round. Causal processing can produce a token with the intrinsic shape of the sentence '*The earth is round*', but not one that intrinsically corresponds to that state of affairs.

What causal processing supplies is a function from intrinsic properties of input tokens to intrinsic properties of output tokens. For this causal function to yield an information function, those properties must be correlated with propositions. The correlation need not be one–to–one: tokens with different intrinsic properties – e.g. upper-case instead of lower-case tokens of 'THE EARTH IS ROUND' – may well be processed in the same way. So the relevant intrinsic properties of tokens need not be a function of the correlated propositions. But the propositions must be a function – a 'semantic function' – of them if the causal function is to embody an information function. A semantic function, in short, is what makes a causal process a computation. Any computer, i.e. any causal system for processing information, must impose or exploit some suitable semantic function.

There are many semantic functions: the same proposition can be processed by many different systems. Think of the range of tokens of the proposition that the earth is round: all the different sentences to that effect in the world's many languages; all the kinds of globes, photographs and other depictions of the earth that are used *inter alia* to show that it is round. But in a single computation this proposition will be some definite semantic function of one or more of its tokens' intrinsic properties: e.g. of any visually recognisable variant of the printed sentence '*The earth is round*'. And those properties in turn define a type, any token of which would, given the same semantic function, be *ipso facto* a token of the same proposition. In any computation using that function, any token of that type will *represent* that state of affairs, the earth's being round.

Representing a state of affairs by intrinsic properties of a token proposition, by which it is causally processed, is the core of computing. But computers represent more than states of affairs. They also represent their constituents: e.g. the earth (or the concept of it), which is a constituent of real or supposed facts about it. They do this by using semantic functions that are compositional, i.e. which represent a state of affairs by a token's stuctural properties: properties that are spatial or temporal functions of properties of its parts, which in turn represent constituents of the state of affairs. Thus the shape of a printed token of '*The earth is round*' is a spatial function of the shapes of the words in it that represent the earth (or the concept of it) and the property or concept of being round; just as the sound of a spoken token of that sentence is a temporal function of the sounds of its spoken words. These further functions we may call 'syntactic', because collectively they constitute a computing system's grammatical rules of composition, its *syntax*; just as the arguments of those functions – e.g. English words – are its syntactic elements, its vocabulary.

The arguments of syntactic functions naturally include phrases, clauses and whole sentences as well as words: grammar covers sentential as well as sub-sentential combinations. '*Mars and the earth are round*' is as much a syntactic function of '*Mars is round*' and '*The earth is round*' as they are of their sub-sentential components. So in particular, a causal function that is a computation is automatically a syntactic function, from the causally relevant types of the input tokens to those of the output token. And the total correlation of a system's syntactic types (the arguments and values of all its syntactic functions) with what those types represent (states of affairs and their constituents) is the system's *semantics*.

course to stop confining them to natural languages. They become features of any system, physical or mental, digital or analogue, linguistic or pictorial, that processes information. But even so, syntax and semantics remain interdependent. In particular, *pace* Stich,[4] a syntax presupposes a semantics. What makes some of a token's intrinsic properties syntactic is that the semantics of some processing system, of which the token is a causal part, makes some state of affairs a function of them. Semantics does not exploit syntactic properties: it creates them. Until a semantics uses it, no intrinsic property, or causal process, is syntactic.

3. So far, I hope, so trite. But what does it take to process information, i.e. to turn a causal process into a computation? That will determine what a computer is, and hence the sense of our question. But this sense must be constrained to make the answer to the question enlightening about the mind. So on the one hand 'information' and related terms must not be so restricted that only minds could process information. Our paradigm, the general purpose programmable electronic digital computer, must not need a mind in order to be a computer. And on the other hand, we must not let too many causal processes count as computations. Otherwise our question will tend toward the trivial: how many mental processes are causal? Probably most of them, whether they are material or not. But that is no news, and does nothing to mark off mental from other causal processes.

This needs saying because the question is easily trivialised in this way, via a sense of 'information' that makes some causal functions seem semantic when they aren't; which then makes other causal functions seem syntactic when they aren't. To see how that can happen, consider first a causal function that really is semantic (as of course most are not, just as most semantic functions are not causal): veridical perception, where the very fact that makes a proposition true is what causes us to believe it.

Suppose for instance I see that one object is longer than another ($a>b$) and the same length as a third ($a=c$). In the circumstances (the objects are clearly visible, my eyesight is good, etc.) I acquire these beliefs only because they are true: the facts $a>b$ and $a=c$ are necessary causes of my believing them. These causal processes here embody semantic functions: the effects are token beliefs '$a>b$' and '$a=c$' which represent the very facts that cause them. But for precisely this reason, the processes do not embody syntactic functions: they are not computations. The effects represent their causes, not some function of what their causes represent – if only because their causes ($a>b$ and $a=c$) represent nothing.

But now suppose I infer from these beliefs that $b>c$. This too is a causal process, the tokens '$a>b$' and '$a=c$' combining to cause a new token belief '$b>c$'. But this process does not embody a semantic function: '$b>c$' does not represent the token beliefs that caused it. What it represents is a function – an information function – of the states of affairs they represent. So the inference, unlike the perceptions, *is* a computation: its causation does embody a syntactic function – which is made syntactic by the semantic function embodied in its perceptual precursors. See Figure 1.

Now compare the situation shown in Figure 1 with one in Newtonian

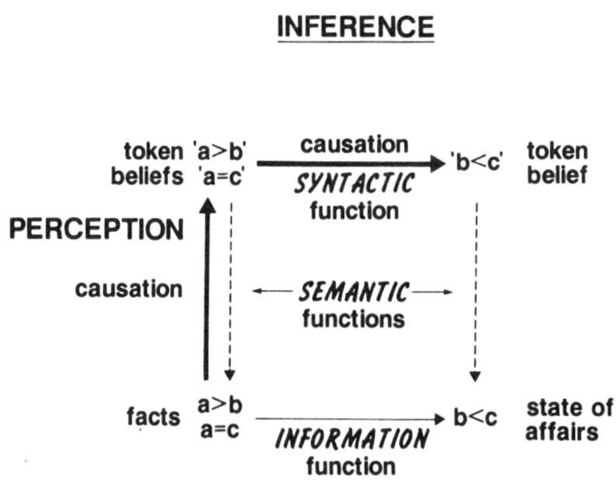

Fig. 1

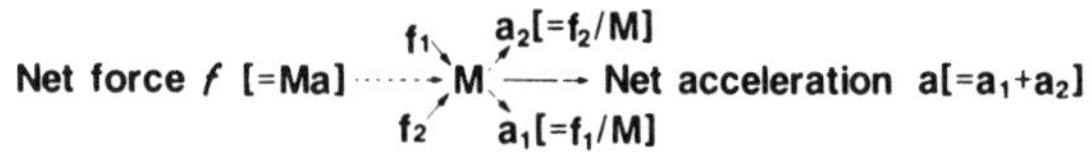

Fig. 2

mechanics that has the same causal structure, but where none of the causal functions is either semantic or syntactic. Suppose two forces f_1 and f_2 act in different directions on a mass M, causing accelerations a_1 and a_2 that combine vectorially into a net acceleration a. Any force F acting on M is a well-known function ($F = MA$) of the acceleration A which it causes, and the notional net force f on m is defined to be the same function of M's net acceleration a. (For brevity I am here using 'f' etc. to refer both to token forces and to their strength; and similarly for 'M', 'a', etc.) See Figure 2.

Now pretend that $F = MA$ is a semantic function, i.e. that the accelerations a_1 and a_2 represent the forces f_1 and f_2 which cause them, and that the net acceleration a represents the net force f. (Strictly speaking: pretend the magnitudes a_1, a_2 and a of these accelerations define types 'f_1', 'f_2' and 'f' of events representing states of affairs in which forces of magnitude f_1, f_2 and f are acting on M.) This makes the vectorial combination of M's accelerations a syntactic function, representing an information function from applied to net forces. In effect, we are pretending that the mass M 'perceives' the forces f_1 and f_2 and then 'infers' a net force f by using vectorial 'addition' to compute an output token a from the input tokens a_1 and a_2 as shown in Figure 3.

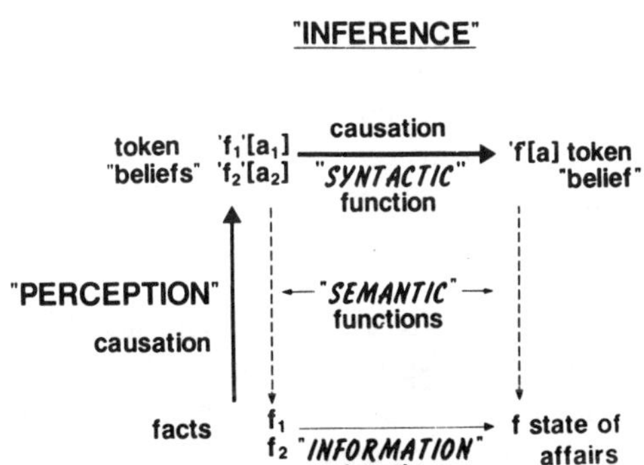

Fig. 3

Figure 3 is of course a fantasy. The accelerations of objects do not represent the forces that cause them, so their vectorial combination is not a computation. But there is a sense of 'information', in which effects embody information about their causes, that is apt to induce this fantasy: the sense used in communication theory.[5] In it, *M*'s acceleration a_1, in direction *d* (say) embodies the information that the force acting on *M* in that direction is f_1, rather than anything else. The information reduces to 1 a greater number of initially possible forces; and the greater the reduction, the more the information. In this sense information is a quantity, increasing with the number of possibilities eliminated. If *N* equally likely forces might have acted on *M* in direction *d*, the acceleration a_1, showing that only f_1 did, embodies $\log_2 N$ 'bits' of information. ('Bits' = 'binary digits', the number of digits in the binary numeral '*N*'.)

But 'bits' of information are not pieces of it. The propositions P_1, and P_2 that forces f_1 and f_2 act on *M*, are pieces of information, however few or many bits they are. But however much information this is, a_1 and a_2 are not tokens of it. They embody it only in a dispositional sense: since they are a causal function of the facts that make P_1 and P_2 true, those facts are what they would represent if that function were semantic.

That sense is good enough for communication theory, which is concerned with how much information causal links could process, whether they actually process it or not. But it is not good enough for us. Our question is whether *M* actually does process information about forces acting on it, not whether it could. And it would only do so if its accelerations represented the forces that caused them, which clearly they do not. Effects do not automatically represent their causes, and nothing else here makes them do so. That is why Figure 3 is fantastic: not because it makes no sense, but because it is plainly false. 'Computational mechanics' might be true: it just isn't. a_1 and a_2 do not actually embody any information, and their vectorial combination does not actually process it.

4. We must take computers to embody the information they process in an actual and not merely a dispositional sense. The concept of computation is after all meant to explain how inference and other mental processes differ from the reactions of masses to forces. We must therefore set aside concepts of information, and hence of computation, that are too weak to mark the difference. (And if we think there really is no

difference, it would be better not to call minds computers at all than to do so in a sense that puts the priciest product of IBM on a par with a simple mass.)

So when *is* a causal process a computation? I said in section 2 that it is when a semantic function makes the causally relevant properties of its stages syntactic. This implies that the semantics of computation is what generates its syntax. But it is not so obvious what generates its semantics: what makes a stage of a causal process actually embody information. So it is worth asking if I have got the interdependence of syntax and semantics the right way round. Perhaps syntax comes first after all. Perhaps information is embodied *because* it is processed, rather than *vice versa*. But if so, there should be a criterion for a causal process being syntactic that is independent of semantics. But is there?

There are two *prima facie* candidates. (1) A process is syntactic when it is governed not by laws but by *rules* (as for instance our use of language is). (2) A process is syntactic when, like a Turing machine, it embodies an *algorithm* that can take input, and yield output, of unlimited complexity. In fact neither candidate will do; but seeing why not will help to show why computation really must be defined semantically.

Take the rule-governed criterion first. Here of course we must not take it psychologically: the rules must not need to be laid down by people. Nor may we appeal to representations in saying what rules are. The only non-question-begging sense of 'rule-governed' seems to be this: a rule-governed system is one that satisfies a generalisation because a sub-system corrects incipient deviations from it. This is certainly a sense in which our use of language is rule-governed: we remain disposed to use words properly by being disposed to correct each other's misuses of them.

(1) does then rule out the accelerating mass M of section 3, which has no sub-system to correct deviations from the proper vectorial combination of its accelerations. But it still admits far too much. Consider how a floor keeps you level as you walk across it. As you move, gravity makes you start to fall. That in turn makes the floor bend beneath you – until its elastic reaction produces an upward force to match your weight and stop your fall. So according to (1) you and the floor are a rule-governed system: you satisfy a generalisation (that you stay on a level) because a sub-system (the floor) corrects incipient deviations from it.

This is nonsense. It makes the corrections real-time computations – which need to be remarkably fast and accurate: too little, too late or in the wrong place, and you fall onto or through the floor; too soon or too much, and you levitate. But you don't: the floor is a remarkable computer! And so, according to (1), is any rigid object, the elasticity of whose surface will preserve its shape against any random bombardment by air molecules. So indeed is any system with equilibrium states made stable by causal mechanisms. Of course we and our computers are such systems, and would never survive if we were not. But then so are all macroscopic objects. (1) is a hopelessly weak criterion for syntactic processes.

Can (2) do better? Not unless 'algorithm' also has a non-question-begging sense, weaker than its dictionary definition ('method or procedure of computation'). But the only suitable sense seems to be 'function' as I am using the term; and then (2) does not even rule out our mass *M*. For in this sense *M* does embody algorithms, such as the vectorial addition function of its accelerations illustrated in Figure 2. And the function is not limited to the two arguments shown there: it can take unlimitedly complex input – e.g. the impact of any number of air molecules. And *M* also embodies functions with unlimitedly complex output – e.g. functions from the momenta of any number of molecules hitting *M* to their momenta immediately afterwards.

These functions are admittedly not embodied in *M* in the way they would be in Turing machines or other computers. Specifically, they are not programmed into *M*. But we cannot save (2) by limiting algorithms to functions that have been programmed. 'Program' is a semantic term: computer programs *represent* functions, i.e. their tokens represent states of affairs in which the functions' arguments yield their prescribed values. To program a computer is to input a token of the program in order to make the computer embody the very function the program represents; and to run the program is then to input tokens of the function's arguments, thereby outputting tokens of its corresponding values. In short, to program a causal function is just to generate it from a representation of it. But what it is to do that depends on what it is to represent something: this criterion for a causal function's being syntactic depends on semantics.

And anyway, the criterion doesn't work. Programming a causal function does not make it syntactic, because it does not make its arguments and values represent anything. Suppose for instance we program *M* to combine accelerations vectorially by giving it an engine that accelerates it at a_0 when tokens of a sentence 'Let $a = a_0+a_1+a_2$'

are input. a is now whatever function of a_1 and a_2 those token sentences represent: *M*'s vectorial combination function has been programmed into it. In particular, the function $a = a_1+a_2$ is programmed into it, by making a representation of that very function turn its engine off. But that does not make *M* a computer: it does not make the combination of a_1 and a_2 syntactic, because it does not make them represent anything, and in particular not the forces that caused them.

In short, though computers can be programmed, that is not what makes them computers. Of course programs are used to induce in computers causal functions that will match their intended semantic functions, and thereby become syntactic counterparts of those functions. And of course every program has a semantics, because it represents the function it induces. But that is not what makes that function syntactic. What makes it syntactic is that its arguments and values represent something; and the program that induces it is not what makes them do that.

The fact is that (2) is credible only if an algorithm's input and output are already taken to be representations; just as (1) is credible only if the sub-system that makes a system rule-governed is taken to react, not to deviations from a generalisation, but to representations of them. But then what it is to be an algorithm, or to be rule-governed, depends on what it is to represent something, not the other way round. The syntax of computers is after all a product of their semantics. Only when we know where the semantics come from will we know what it takes to be a computer, and how much of the mind may be one.

5. It is in fact obvious, once we think of it, where the semantics of computers come from: they come from us. The computers that have prompted and given sense to our question compute because we compute with them. Computers may in future compute for themselves. But at present they compute for us, and represent what we use them to represent.

When do we compute? First, when we infer, i.e. derive new beliefs from old ones. Whatever computing is, it includes inference; and whatever representations are, they include token beliefs. How beliefs represent states of affairs is another question, which I fortunately need not answer. All I need is the fact that they do. My believing any proposition *P* embodies the information *P*; and any inference made from that belief processes that information. The inference of Figure 1,

COMPUTER-AIDED INFERENCE

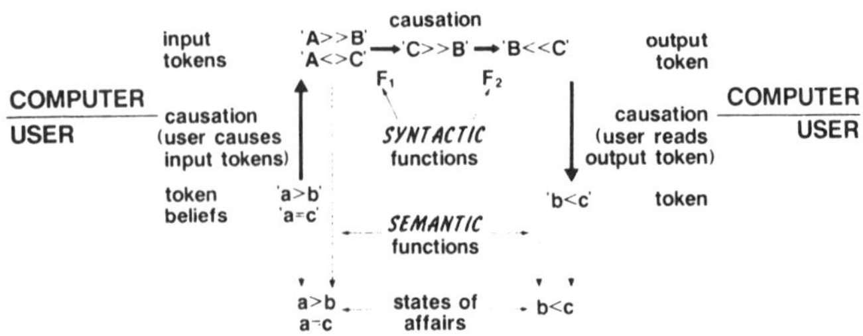

Fig. 4

where my token beliefs '$a>b$' and '$a=c$' cause a token belief '$b<c$', is a paradigm of computation.

To see how this paradigm can bestow its semantics on a computer, let me use one to make the inference of Figure 1. Suppose I give my computer language a compositional semantics, with symbols 'X', 'Y' . . . representing objects, and '$>>$', '$<>$' and '$<<$' representing the $>$, $=$ and $<$ relations. That is, to make inferences from what I take to be facts of the form $x>y$, $x=y$ or $x<y$, I input symbols of the form '$X>>Y$', '$X<>Y$' or '$X<<Y$'; and I let output of those forms cause me to acquire the corresponding beliefs.

So now my token beliefs '$a>b$' and '$a=c$' make me input tokens '$A>>B$' and '$A<>C$'. Suppose they are processed in two steps. First one function (F_1) makes them cause an internal token '$C>>B$'; and then another (F_2) makes that cause an output token '$B<<C$'. Finally my seeing the output completes the inference by inducing in me the token belief '$b<c$'. All this is shown in Figure 4.

The input and output tokens here are clearly representations, and would be even if I weren't computing with them. My computer language is a language like any other: its tokens are of types that regularly cause, or are caused by, beliefs in propositions they thereby embody, as they do here. '$A<<B$', '$A<>C$' and '$B<<C$' represent the states of affairs $a>b$, $a=c$ and $b<c$ just as tokens of any other language would do.

And the internal token '$C>>B$' represents the state of affairs $c>b$ (which may or may not be the same as $b<c$). Of course, being internal, its intrinsic properties differ from those I directly cause or perceive. But that is immaterial: computers work with electronic rather than with

printed tokens only to speed things up. Their syntactic functions are the same whatever intrinsic properties they work with. And though the token '$C>>B$' is neither an effect nor a cause of a corresponding belief in me, it is related by syntactic functions to tokens that are. Take F_2, and suppose it turns '$C>>B$' into the output '$B<<C$' by replacing the token '$>>$' with a token '$<<$' and swapping the tokens 'B' and 'C'. But these tokens represent *b* and *c* both in the output and in the input that causes '$C>>B$', just as '$>>$' represents $>$ and '$<<$' represents $<$. But '$C>>B$' might have been the input: I might have used the computer to infer $b<c$ from $c>b$. Then '$C>>B$' would certainly represent $c>b$. But F_2 has the same semantics whether it is the whole or (as here) only part of a computation. So '$C>>B$' represents $c>$b anyway.

It also represents it when I use the computer to infer $b<c$ conditionally, i.e. without believing in it, because I don't believe in $a>b$ and $a=c$. That inference too is a computation, though the mental token '$b<c$' it causes is not a belief. But it does make me believe that *if* $a>b$ and $a=c$, then $b<c$: i.e. believing those two propositions would now make me believe that one. In short, a conditional inference just disposes me to make the unconditional one: my mental tokens are processed *as if* they were beliefs. The semantics of conditional computation is that of its unconditional counterpart. In particular, F_1 and F_2 will have the same semantics in a conditional inference as in an unconditional one; and so therefore will '$C>>B$'.

Computers are widely used to make conditional inferences (e.g. to predict consequences of alternative economic policies), whose input and output are thus not to be unconditionally believed. This tends to obscure their semantic dependence on unconditional inference and hence on belief. It also explains why their input and output seem sometimes not to represent states of affairs at all. Suppose for instance I make my computer work out squares by programming the x^2 function into it. Suppose I then input a token '5': it outputs '25'. These input and output tokens are numerals, not sentences, representing respectively the numbers 5 and 25. And numbers, whatever they are, are not states of affairs: they correspond to no propositions, true or false. One cannot believe or disbelieve 5 or 25 or any other number.

Yet I do process information when I compute my squares. I am in fact using my computer to make a general conditional inference: if the number of anything is 5, its square will be 25. That inference disposes me, if (e.g.) I believe that the side of a square is 5 cm, to believe also that its area will be 25 cm^2. And if I were not using the computer to make such an inference, its input and output tokens would not even be numerals. On their own they can no more represent numbers than the token '*C*>>*B*' can represent the state of affairs *c*>*b*. They, like it, represent only what the inferences they are part of make them represent.

6. All this, however obvious, may well seem to give our question little useful sense. That computers compute because we infer with them tells us more about them than it does about the mind. Calling inference 'computation' makes that aspect of the mind a computer only by definition. Still, there can be informative computational accounts of *how* we infer.

Suppose for instance I make the inference of Figure 1 without using my computer. The process is still causal, only now it is inside me. Suppose again it is a two-step process, as shown in Figure 5, and an unconditional inference, i.e. its 'input' and 'output' tokens are beliefs.

The two steps of the process may be inferences: the intermediate token '*c*>*b*' may be a belief. But it may not be conscious: I may not know how I infer '*b*<*c*'. That gives scope for a computational theory, which may be tested (e.g.) by how fast various inferences are made: roughly, the fewer steps the faster. Thus if this one is as shown, it should be slower than a one-step inference to '*c*>*b*'; and that is a testable prediction.

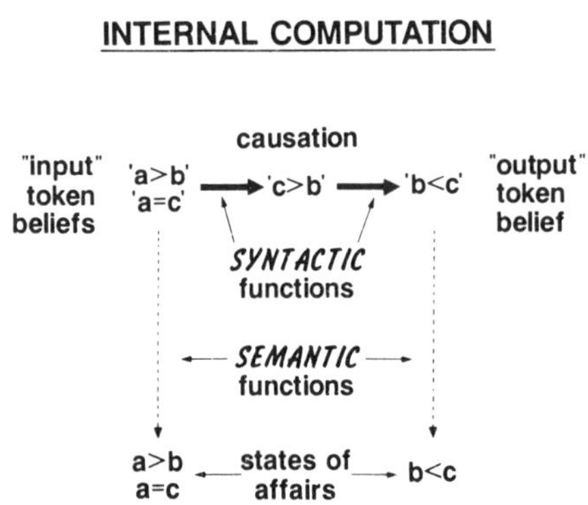

Fig. 5

And such tests do support computational theories, like Johnson-Laird's[6] theory of syllogistic inference. He says we do syllogisms using 'mental models' of premises like 'All the *A*s are *B*s', i.e. with mental arrays like:

$$
\begin{array}{l}
\text{`}a = b \\
\;\;a = b \\
\qquad\;\;(b) \\
\qquad\;\;(b)\text{'}
\end{array}
$$

in which token '*a*'s and '*b*'s represent arbitrary *A*s and *B*s and '(*b*)'s represent possible *B*s that are not *A*s. Since according to the theory some syllogisms need more models than others, they should take longer; and Johnson-Laird cites evidence that they do.

So there would be scope for computational psychology even if mental computation were restricted to inference. And it isn't, not even within inferences. The two steps of the inference in Figure 5 may be computations without themselves being inferences, because the token '$c>b$' may not be a belief: not because these steps are only conditional inferences (they aren't), but because '$c>b$' may not combine with desires to cause action in the way beliefs do.[7] If '$c>b$' is a belief and (for instance) I want something longer than *b* when only *c* is to hand, it will make me take *c*. But it may not: whatever I want, it may only make me believe $b<c$. But then it is not a belief, and wouldn't on its own represent $c>b$ at all. (Why not is a question I cannot tackle here.) Like my computer token '$C>>B$', it represents $c>b$ only via the token beliefs that are its causes and effects. But it does then represent it, as '$C>>B$' does. The steps of my inference may not be inferences, but they are still computations. Computation can outstrip inference even when it gets its sense from it: the computational psychology of inference is not a trivial subject.

7. Nor, I admit, is the computational psychology of perception, in which non-inferential computing may also occur, both inside and outside the perceiver. That it occurs outside, in indirect perception, is undeniable. Suppose for instance I perceive a temperature *T* indirectly by using a thermometer, e.g. one with a visible volume *V* of mercury that varies with *T*. (*V* represents *T* by being marked in degrees Celsius, so that it induces beliefs about *T* rather than about *V* itself.) Now

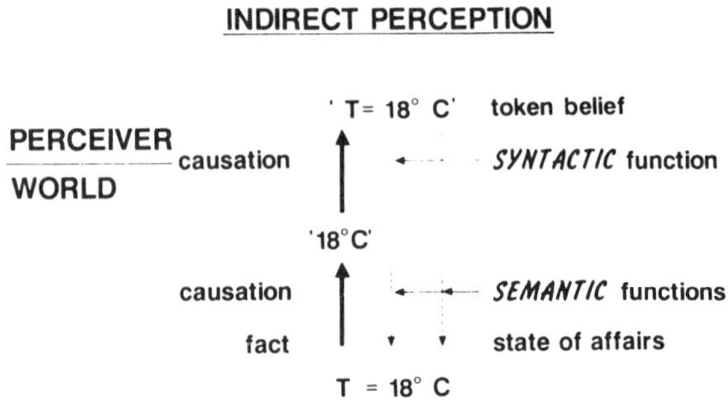

Fig. 6

suppose in particular it tells me that *T* is 18° C. It does so in two steps (analogous to the steps of the inference of Figure 4). First, *T* makes it read '18° C'; and second, my seeing that token causes in me a token belief '*T*=18° C'. See Figure 6.

In this process of indirect perception, the thermometer reading '18° C' represents what the belief it causes represents. But that is the very fact *T*=18° C, which has caused it. The cause here is a semantic function of its effect, not a syntactic one. '18° C' represents *T*=18° C, not a function of anything that fact represents (it represents nothing). The first step of the process is not a computation.

The second step is. My belief '*T*=18° C' represents, not the token '18° C' that causes it, but a function (the identity function!) of what that token represents, namely *T*=18° C. My belief is the output of a computation, albeit one embodying a trivial information function (identity).

But perceptual computing need not be trivial. Suppose a photograph tells me that a man has been running (*R*). What the picture, being still, actually represents is his being in a running posture (*U*). But though I react to the depiction of the state of affairs *U*, I may not notice it. I may only notice, i.e. be caused to believe in, *R*. I cannot then have inferred *R* from *U*, since I never believed in *U*. But I still believe in *R* because the picture represents *U*: my token belief '*R*' is computed from the pictorial representation '*U*'.

In both these cases the computing starts outside me. But in perception, as in inference, it may also start inside; and that makes it the province of computational psychology. Thus Marr[8] for instance says that vision starts with 'arrays of image intensity values, as detected by

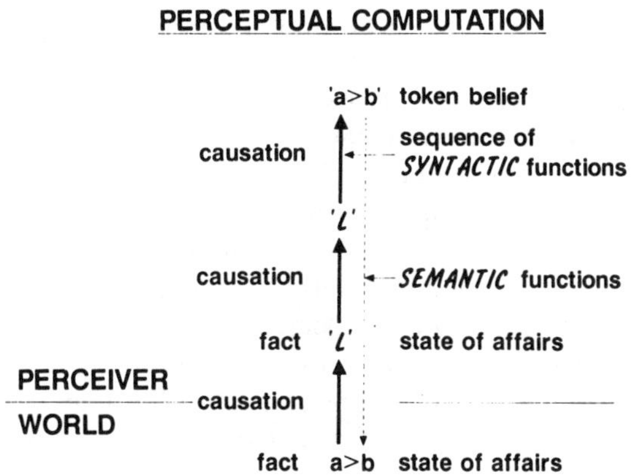

Fig. 7

photoreceptors in the retina' and proceeds 'by mapping from one representation to another' (p. 31) until it causes a belief about 'what is present in the world and where it is' (p. 3).

So according to Marr, when I see that $a>b$, light reflected from a and b causes an array L of light intensities to fall on photoreceptors in my retina. Their reaction to L he says represents it, and a sequence of computations then transforms this 'viewer-centred' representation (via e.g. representations of a's and b's edge-orientations) into the 'object-centred' belief '$a>b$' as shown in Figure 7.

Marr's representations start in the retina because that is where, in direct visual perception, systematic links with belief start. The optic nerve embodies definite causal functions, which transmit the semantics of visual beliefs back to some of their causes. (In indirect perception we extend the links outward to thermometers, photographs, etc., thus letting them represent states of affairs also. But that is not psychology: the mind starts – if only by definition – at the body's boundary.)

Marr's theory certainly makes sense. If thermometers and photographs can embody propositions by systematically making us believe related ones, so can our sense organs. If doing that can make $T=18°$ C's causation of '18° C' semantic, why not L's causation of 'L'? If it can make the causal function from 'U' to 'R' syntactic, why not the function from 'L' to '$a>b$'? After all, the function need not embody an inference: it need not make perception deliver even unconscious beliefs about our retinas. '*L*' need be no more like a belief than a thermometer reading or a photograph is.

And the theory is testable. It postulates specific sequences of computations, from tokens like '*L*' to the beliefs they cause, whose syntactic functions must be embodied causally in the optic nerve. If corresponding causal functions are found, then so far so good; if not, the theory must be altered or abandoned. It is an empirical matter whether, how much and how our eyes – and other sense organs – compute.

How much computing there is in perception is therefore not for me to say: maybe a little, but maybe a lot. And inferring is computing by definition. But there is much more to mental life than perceiving and inferring; and none of the rest of it, I shall now argue, is computation.

8. Many mental processes are not computations because their stages are pains or other sensations, which represent nothing. If for instance a loud aural sensation *S* gives me a headache *H*, that is a mental process. But only the tokens *S* and *H* are processed: no information is, because nothing is represented. True, *S* and *H* are self-intimating (when they occur, they make me believe it), so the process produces a parallel sequence of beliefs, which are representations; but it is not a computation either. My token belief '*H*' (that I have *H*) is not inferred from my belief '*S*' (that I have *S*): no causal process leads from '*S*' to '*H*'. '*S*' is a mere side-effect of *S*. The causal order is not '*S*' → *S* → *H* → '*H*', but that shown in Figure 8.

No causal function in Figure 8 is syntactic. The functions from *S* to '*S*' and *H* to '*H*' are semantic, not syntactic; and the function from *S* to *H* is

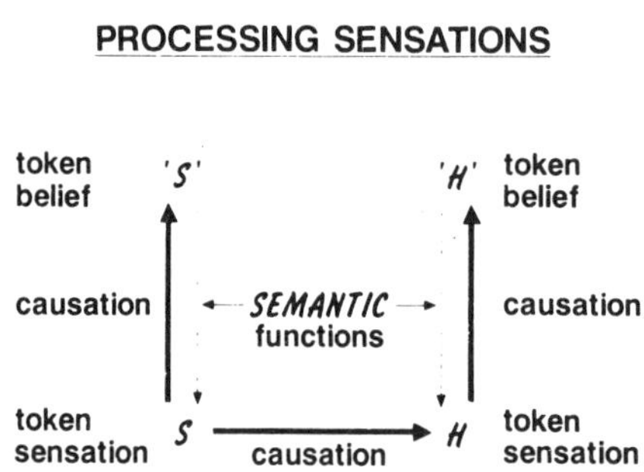

Fig. 8

neither, since *S* and *H* represent neither each other nor anything else. There is no computing in the processing of sensations.

But the mind processes more than sensations on the one hand and beliefs on the other. It processes other propositional attitudes, and they too are representations in their own right. If believing in *P* represents *P*, so do wanting, hoping for and fearing *P*. And all these attitudes figure in causal processes that depend on the propositions they embody. Suppose for instance I close a door, i.e. make a state of affairs *C* (the door's being closed) obtain, because I want less of a draught (*D*) and believe that if *C* then *D* ($C \rightarrow D$). My action and my desire are tokens respectively of the propositions '*C*' and '*D*', as my belief is of '$C \rightarrow D$'. My action embodies a 'proposition function' of that desire and that belief, just as an inference, such as the inference of Figure 1, does of the beliefs it starts from.

And as in inference, this function must have a syntactic surrogate. It cannot be given just by the causation involved. Causal processes, I noted in section 2, work on intrinsic properties, and embodying a proposition is never an intrinsic property. So my token desire, like my token belief, must have some causally relevant intrinsic property of which *B* is a semantic function: i.e., a syntactic property. And as for desire, so for every propositional attitude. The causal processing of propositions embodied in tokens of all attitudes must be syntactic. But why then call it 'computation' only when the attitude is belief (or, as mentioned in section 5, is being processed, in a conditional computation, as if it were a belief – a qualification I hereafter take as read). What makes the processing of belief so special?

The short answer is that processing tokens of other attitudes is not processing *information*. Pieces of information are all propositions: but not all propositional attitudes embody information. Only beliefs do. Believing in *P* does embody the information that *P*: wanting, hoping for and fearing *P* do not. That is why inferring is processing information, i.e. computing, and the processing of desires, hopes and fears is not.

But why do propositional attitudes only embody information when they are (or entail) beliefs? To see why, consider first why questions and commands do not embody information. The spoken question '*Is the door closed?*' ('*C*?') is a token proposition: it represents the state of affairs C; but no one would say it embodied information. Nor would one say that of the command '*Close the door*' ('*C*!'), which also represents *C*. And the reason is obvious: these tokens don't represent *C* as obtaining, i.e. as a fact: they don't embody the proposition '*C*' as true. '*C*?' obviously doesn't, since it is asking whether '*C*' *is* true. And nor does '*C*!', since when '*C*' is true (when the door is closed) the command is redundant.

Information, in short, takes states of affairs as facts: tokens of information embody propositions as true. They may of course not be true: information can be false. But it is meant to be true; and is processed accordingly, even in conditional computing. Information functions are always meant to preserve truth: the reliability of all computing is measured by how rarely it turns true input into false output.

Desires, hopes and fears do not embody propositions as true. A desire for *P* does not take *P* for a fact; nor does a hope or fear of *P*. That is why those attitudes do not embody the information that *P*. A belief in *P* does take it for a fact, i.e. embody '*P*' as true. To take *P* for a fact just *is* to believe it, whether or not one has any desire, hope, fear or any other attitude towards it. That is the defining characteristic of belief (and why there is the so-called 'paradox of belief':[9] the essential absurdity of saying of any *P*, '*P* [is true] but I don't believe it' – a paradox that has no analogue for any other attitude).

The reason inference and perception aim for truth is that what they deliver is belief. That is why inference is by definition computation, and how inference and perception can include computations that are not inferences. My reaction to light falling on my retina embodies information (if it does) because it is then processed by functions that generally preserve truth in order in the end to yield a true belief.

Other attitudes often do not aim at truth at all, and then there is no need to preserve it. Sometimes falsity is preserved: e.g. in processing desires, which take states of affairs *not* to be facts. (We don't want what we already have: we may want to keep it, but that's another matter.) Take my closing a door because I want less of a draught. The causing of that embodies a general proposition function <'*Q*', '*P*→*Q*'→'*P*', where '*Q*' is a desire, '*P*→*Q*' a belief and *P* the state of affairs brought about by the action they cause. The function is a common one because it always transmits the unwanted falsity of '*Q*' to '*P*', which the action then remedies. It is not (one hopes) so common when '*Q*', '*P*→*Q*' and '*P*' are all beliefs, since it notoriously does not always transmit truth: it is not a good information function.

Conversely, my wanting two states of affairs together, *P*&*Q*, will often not make me want *P*, because I want only *Q*, or only *P*&*Q*, because I have *P* already. That is, the function '*P*&*Q*' → '*P*' is applied much less often to desires than it is to beliefs partly because, though it always transmits truth, it doesn't always transmit falsity. It is a good information function, but not a good function for desire.

As for desire, so for other attitudes. Proposition functions vary with the attitudes embodied in their arguments and values: functions common for some attitudes will be rare for others. Now to say of what causal functions any attitude's tokens are arguments or values is to say how they are caused, how they interact and how they affect our behaviour. And that, on a functionalist view of mind,[10] is enough to define them. But even if it isn't, it still tells us a lot about them, and hence about the mind. So discovering these functions must be a large part of psychology, if it is to explain our propositional attitudes (as opposed to explaining them away). To say what they are, and why we have them, it must say how they affect our processing of propositions, i.e. what their typical proposition functions are, and why we embody those functions.

It should be obvious then that there is no way computational theories, restricted to truth-preserving information functions typical only of belief, will tell us what we need to know about other propositional attitudes. But in fact two ways are still constantly being attempted, and I should therefore show explicitly why they do not work.

One way is exemplified by a decision theory model of my action '*C*' above, in which I compute expected utilities (EUs) of *C* and not-*C* from my beliefs and desires, and then act to make whichever has the higher EU a fact.[11] The details don't matter: what matters is the idea that I form token beliefs like 'I want *D*' and 'I believe $C \rightarrow D$' and apply decision theory to them to compute a token 'I should do *C*'. Like the computational mechanics of section 3, the idea makes perfectly good sense: only it's obviously false. I *could* act by using decision theory to work out how to act and then acting that way. But I don't; any more than a mass *M* uses Newton's laws to work out how to react to applied forces. It simply reacts as the laws say it will. Similarly, my beliefs and desires may well cause me to act as Jeffrey's theory says they do. That doesn't mean I compute my action from them, and I don't.

The objection to such computational theories is not that they are nonsense, or vacuous, but that they are false. Unfortunately in psychology they are also tempting. No one thinks the truth of a computable theory of mechanics makes *M* a computer, because we know *M* can't compute. But we can: hence the tempting idea that what makes a computable theory of the mind true is that we compute with it. And precisely because this *might* be true, we must not stretch the meaning of 'computation' to make it true by definition. That will only make M a computer too, and evacuate real computational theories of the mind of their substantive content.

The other way of making our processing of other propositional attitudes look like computations is to identify them with beliefs in related states of affairs. Thus we might say that wanting P is really believing P to be desirable; that fearing it is believing it to be fearful; and so on. That is, for each attitude A we find – or define – a propositional operator O_A such that $A(P) = B(O_A(P))$ for all P, where B is belief (and O_B is the identity operator). This transforms all A's proposition functions into corresponding ones typical of belief, i.e. into truth-preserving information functions. We make all our processing of propositions computational in appearance.

But not in reality. Calling a desire for P a belief, that P is desirable, may (by definition) make processing preserve the truth of 'P is desirable'. But not the truth of 'P'. The proposition functions in which 'P' figures of as a token of an attitude A remain what they always were. And we shall not discover what functions they are by transforming them into information functions of '$O_A(P)$'. On the contrary, until we know what functions they are, we shall not know how to transform them, i.e. what O_A is. This way of making the mind look like a computer is a trick: it tells us nothing about the mind.

Both these ways of making theories of all propositional attitudes computational fail; and I see no other way. I conclude that the psychology of attitudes other than belief will never be computational in any serious sense. Computational psychology will remain restricted to the aetiology of belief, especially in perception and inference. That part of the mind may well be a computer; the rest of it, which is most of it, is not.

NOTES

* This paper is a sequel to my reply[12] to Margaret Boden,[13] stimulated by criticisms made at a symposium, and later by Philip Johnson-Laird and Jerry Fodor. It was developed through seminars at the Institute for Systems Studies in Moscow and at La Trobe University before being given to the Sydney Symposium on Cognitive Science in August 1985. Since then it has been further revised in the light of seminars at the Queen's University Belfast, Trinity College Dublin, the Institute of Philosophy in Newcastle-upon-Tyne, Cambridge University and St David's University College Lampeter. I am much indebted to those who took part in all those seminars, to others at Adelaide

University and the Australian National University with whom I discussed it, and to Eliott Sober, Huw Price, Peter Gibbins, Benedict Heal and Jeremy Butterfield, whose comments on this and related topics have influenced it (if not always in ways they would approve of).

[1] M.A. Boden (1984) 'What is Computational Psychology?', *Aristotelian Society Supplementary Volume* **LVIII**, pp. 17–35.

[2] S.P. Stich (1983) *From Folk Psychology to Cognitive Science*, MIT.

[3] J.A. Fodor (1981) *Representations*, Harvester.

[4] See *op. cit.* (note 2).

[5] F.I. Dretske (1981) *Knowledge and the Flow of Information*, Blackwell.

[6] P.N. Johnson-Laird (1983) *Mental Modes*, Cambridge University Press.

[7] D.H. Mellor (1984) 'What is Computational Psychology? **II**', *Aritotelian Society Supplementary Volume* **LVIII**, pp. 37–53 (at p. 43).

[8] D. Marr (1982) *Vision*, Freeman.

[9] G.E. Moore (1942) *A Reply to My Critics – The Philosophy of G.E. Moore*, P.A. Schilpp (ed.), Northwestern University Press, pp. 553–677.

[10] N. Block (ed.), (1980) *Readings in the Philosophy of Psychology* **I**, Methuen.

[11] R.C. Jeffrey (1983) *The Logic of Decision* (second edition), University of Chicago Press.

[12] See *op. cit.* (note 7).

[13] See *op. cit.* (note 1).

Faculty of Philosophy
Cambridge University, U.K.

KIM STERELNY

COMPUTATIONAL FUNCTIONAL PSYCHOLOGY: PROBLEMS AND PROSPECTS

1. INTRODUCTION

The last fifteen years has seen the development of two independent but co-evolving ideas: (1) the idea that the mind can be *functionally decomposed*, and (2) the idea that cognition is computation.

1.1 Functional Decomposition

Let me begin with the decomposition of the mind. That strategy has three components.

(i) The first is the idea of *functionalism* itself. Functionalists hold that the essence of a psychological state is what it does: its role in our psychological economy. What makes some neural state of mine (say) the desire for a bottle of Cabernet Sauvignon? Not the *intrinsic* character of that state (an ineffable longing), but its role in guiding action in concert with others of my psychological states. Thus perceptions, memories, intentional states, (more controversially) sensations are *functionally* defined: it's what they do that counts, not what they are made of.

(ii) The second component is the view that the mind is modular. Minds are ensembles of discrete and relatively special purpose modules: our general and flexible intelligence emerges out of the cooperation of lesser and more specialised intelligences that compose us: dim because monomanic homunculi. Plausible modules are the various perceptual systems; long and short term memories, the language faculty, visual and spatial imagination, and so on. This idea of breaking a complex function into interacting less complex functions is of course central to both cognitive psychology and AI.

The modularity thesis has two forms in the contemporary literature. The one just elaborated is Dennett's and Lycan's. Fodor[1] has elaborated a slightly different view: (a) He takes his modules to be autonomous, in that each has relatively limited access to the information in the others. (b) He takes the structure of the mind to be largely *vertical* rather than *horizontal* i.e. instead of there being general memory, or general

Peter Slezak and W.R. Albury (eds.), Computers, Brains and Minds, 71–93

problem solving mechanisms there are systems specialized to particular domains: vision, speech, or whatever. (c) Fodor does not develop a homuncular organization of central congnitive processes, but rather of peripherals. He regards the attempt to construct a theory of central processing as fairly hopeless.

(iii) The final, and vital component is to apply the first two ideas recursively. That is each of the modules is *in turn* seen as an ensemble of more specialized and hence simpler modules. And so on until we reach a level where the functions are so simple and specialized that they are psychologically primitive: they are the equivalents in our heads of the logic gates in a computer. Thus, in psychological theories of visual perception, (perhaps) edge and intensity detectors in our visual cortex might be primitive. To explain their workings we need to refer to neurobiology, not psychology. These are the three elements of *homuncular functionalism*. There is a lot to be said for it.[2]

Firstly, homuncular functionalism enables us to use informational notions in characterising interactions between our mental modules without vicious regress. Our homunculi interact – according to the picture I am advertising – by exchanging information. Perception, for instance, is alleged to be an inferential system aiming at our best hypothesis about the distal sources of our various proximal stimulations, stimulations which are always ambiguous about their sources. In establishing this hypothesis, there will need to be a rich and intelligent exchange of information between the homunculus running (say) Visual Information Incorporated and memory. But that might seem simply circular. Psychology is supposed to *explain* intelligence; but this story about perception seems to invoke it.[3] It's always been an objection to overtly or covertly homuncular theories in psychology that they presuppose what they are meant to explain. The recursive element of the programme of functional decomposition allows us to avoid this objection. We are explaining a clever homunculus (ourselves) by appeal to a group of cooperating but *less clever* homunculi. In turn, each of them is explained by appeal to a group of interacting but even stupider homunculi, and so on, until we reach a 'mere mechanism'. We explain the behaviour of an intelligent being by seeing it as an instantiation of a flow chart. Each box on that chart is a homunculus whose behaviour we can in turn explain by seeing *it* as an instantiation of a flow chart, and so on. We make progress as long as the abilities of the component black boxes are simpler than the abilities they jointly compose.

Secondly, the modular idea is evolutionarily plausible. Intelligence surely did not emerge from the unintelligent in a single step. Less obviously, but still plausibly, it's not just a quantitative growth in the number of neurones. The difference between a sea slug with about 10,000 neurones and us, with about 100 billion, isn't just quantitative. Various new capacities have been plugged into the system, as well as old capacities being extended. If our minds are complexes in this way, we can see how they could have evolved, so long as the components out of which they are constructed are *individually* adaptive. Perception, memory, learning capacities and the like are useful even without our full system being in place.

Finally, this picture is suitably abstract. Early functionalists recognised that psychological descriptions abstracted away from the details of their neural implementation. Any given psychological state – say my desire for a good beef curry – is some neurophysiological state or other. But we don't want to identify the *property* of wanting a good beef curry with a *particular* neurophysiological property. For given the very considerable flexibility of our neural organisation, it's not at all likely that the neural state that implements that desire in me is one and the same as that doing it in another. There's no beef curry neuron, that's turned on when its bearer wants a beef curry, off otherwise.

That gives us a motivation for identifying the psychological property with a functional property. For one and the same functional property could be, and probably is, differently implemented in us.

So early functionalist theorists proposed a dualism of description if not of stuff. We can describe the brain and its doings in abstract, functional, programme-level talk. If we do that, we're talking psychology. Or we can (in principle) describe it in meaty, hardware (wetware) talk. If so, we are doing neuroscience.

Doubtless this was a good idea. But it's too simple. For not all psychological descriptions are equally abstract, equally independent of the meat. Some are highly abstract. We comfortably ascribe intentional and perceptual states to nonhuman animals, whose brains might be quite unlike ours. It's easy to image situations where we would ascribe them to aliens and robots.

But not so others. Take, for instance, sexual desire. It's not obvious, to put in mildly, that sexual-desire-in-humans is psychologically at all like that state in animals; it is for instance much less closely tied to reproduction. Other animals mate only when the female is in season. It is still less clear that this state is one we could share with robots: C3PR is highly camp, but not I think moved by generally erotic inclinations. The same could be said for many other psychological states both of us and of animals.

So we need not two layers of description, functional and structural, but many. Some states – sexual desire, sensations, perhaps some perceptions, will be closely tied to their biological implementations. Others will float relatively free of those implementations: these are tie states that we could share with those of differing biology, or no biology at all. One considerable virtue of homuncular functionalism[4] is that it provides a multiplicity of levels of psychological description.

1.2. Cognition is Computation

The functional decomposition of the mental is married to, or at least cohabiting with, the identification of cognition with computation. It seems obvious that whatever else our minds do, they store and manipulate inner representations: inner states that code information about us and our environment. Minds may not be just information processors, but they are at least that. But what is the *character* of this information processing? One hypothesis – defended most convincingly at M.I.T. – is to identify cognition with computation. We process information in the same way computers do. There is lots to be said in favour of this approach.

(i) It's good research strategy to try to model our information processing on something we already know a bit about. And we do know a good deal about computation, both from the theory of formalised systems of reasoning and from the actual implementation of some of those systems in real machines.

(ii) To some extent, this approach has worked, both in cognitive psychology and artificial intelligence. In particular, there are quite sophisticated models – based on this approach – of visual perception and visual imagination.[5] To the extent you are impressed by the achievements of cognitive psychology, you will take the identification to suggest a promising research strategy.

(iii) One constraint on any theory of cognition is that cognitive processes must be physically realizable. Cognizing is something our brains do, so our theories of cognition must not attribute capacities our brains don't have. A computational theory of cognition meets this constraint. Computation is a fully formalized form of reasoning; any such form is, in principle mechanically realizable.

(iv) There is another, less obvious, constraint on theories of cognition. Cognition involves some set of operations/transformations on states of our central nervous system, states we take to be information bearing. Now:

(a) These states carry information/content in virtue of their *relational* properties, i.e. in virtue of some kind of causal/nomic connection between them and their environment. Thus some state of the visual area of my neocortex is seeing Chomsky only if it's appropriately caused by Chomsky himself.
(b) The processes that access and transform that state are keyed *not* to these relational properties of the central nervous system but to their intrinsic features.

This last remark could use some elaboration. We probably have brain functions specialised for face recognition. Imagine that that includes some device that reads my visual cortex. That reader has no access to the distal causes of the stimulated state of my visual system. If it is to work at all[6] it must be through being honed to the relevant intrinsic features of the complex spiking patterns of my visual apparatus, for that is its only input. Cognition is information processing, but cognitive operations must be tuned to the physical, intrinsic, features of the states that *code* the information, for they have no direct access to the distal causes of those intrinsic features.

So, in some sense, cognition is a *formal* mode of information processing. Its operations are specified independently of the meanings of the formulae on which it operates. Now, computation has precisely this characteristic. In constructing a computational model of some form of reasoning, both the formulae that are the subjects of inferential processes, and the allowed inferences themselves, are specified independently of their semantics. This is the most substantive theoretical reason for identifying cognition with computation.

2. FUNCTIONALISM AND FOLK PSYCHOLOGY

Some version of functional psychology is bound to be right. For the mind has a vast range of structured capacities: perceptual systems, language, a great range of cognitive abilities and cognitively guided motor abilities. I know of no holism so extreme as to deny that this range of capacities doesn't reflect *some* internal organization; the mind has some joints along which it can be carved. The 'Zen Holists' (Dennett's nice, if malicious phrase) don't deny this, nor do the neurophilic Churchlands. They deny only that the organisation of the mind maps, with reasonable accuracy, onto categories identified by folk theory.

The view that the mind is segmented doesn't take us far. It's compatible with a radical rejection of folk psychology and cognitive science, i.e. it's compatible with rejecting the view that a functional theory of the mind will smoothly reduce high level cognitive capacities to brainware via a continuum of intermediate levels. That's the Dennett-Lycan story, but Dennett himself points to the possibility[7] that the capacities most salient to us as persons 'emerge' out of the interaction of a large number of low level functional units.

I have already mentioned two considerations against this concessive move of Dennett's; a third is worth exploring. Homuncular functionalism seems to open the door to an integration of a number of different perspectives on psychology. That picture envisages a hierarchy from the abstractly functional to the wetly structural. Evolutionary biology and the neurosciences fill in the structural pictures; cognitive psychology the functional. But this integrating approach leaves us with a problem child: belief-desire (or folk) psychology. What of the idea of humans as self-directed agents? Where does it fit in? It seems to have only one possible niche. Belief-desire psychology had better be, in some purged form, part of an abstract characterisation of our functional organization. But this leads to a serious problem. Paul and Patricia Churchland have advanced two theses about folk psychology.[8] First, that it is likely to be radically false. Second, that its falsehood is likely to infect cognitive psychology, as that psychology borrows its categories, explicitly or implicitly, from folk psychology.

Two replies are possible. First, the empirical inadequacies of folk theory are somewhat overstated. Intentional psychology (Paul Churchland says) is stagnant and limited. We are no better at explaining people's behaviour in terms of their beliefs and desires than – say – the ancient Greeks. It's severely limited in that there is a huge range of psychological phenomena on which intentional psychology simply has nothing to say: the mechanisms underlying perceptions and learning, insanity, the need for sleep and dreaming, most of our problem solving capacities, and so on.

These *are* serious worries, though Churchland overstates them. Folk theory is *not* as stagnant as he would have us believe: our discoveries about nature affect our folk theory. Our self conception has been profoundly affected by Freudian, Marxist, and probably Darwinian ideas. The ideas of unconscious motivation and belief; the view that beliefs, desires and self-perceptions are rooted in contingent cultural and social histories; the idea that we are biologically driven certainly existed before the nineteenth century. But surely these notions play a much more overt and central role in folk psychology as the result of the intellectual revolutions of the last few hundred years. The claim of stagnation is much exaggerated.

There is no doubt that folk theory is decidedly limited. But on the multiple layer perspective I have been urging we would expect *any* theory to be limited. Folk theory – if it's viable at all – is a theory on the functional end of the functional-to-structural continuum. Its failure to explain – say sleep or insanity – may just show that the explanation for these phenomena is rooted in our biology. Though psychological they, like pain or lust, are to be explained by psychological theories incorporating many of the contingent features of our biological organisation.

A second, and I think stronger, response is to deny the infection theses. It is far from obvious that a rejection of folk psychology should lead us to a rejection of highly abstract functionalist, top down models of (some aspects of) our psychology.

Indeed, there is an argument to show that folk psychology is independent of cognitive psychology; its range of concepts, and domain of explanation, is *not* that of cognitive psychology. For folk psychology is *semantic* psychology. Our thoughts – beliefs, hopes, intentions, expectations – are *about* the world, or aspects of it. They represent, or misrepresent, our environment. Some are about the bomb, others ice cream, still others wine. And so on. Thoughts have content.

Moreover, this is no small deal. For according to folk theory, intentional behaviour is explained by our thoughts in virtue of the *content* of those thoughts. Why did the Dark Lord vaporize New Zealand? Because he did not *want* lamb chops for breakfast again, and because he *believed* that lamb shops come only from an (unvaporized) New Zealand. Our thoughts explain our behaviour in virtue of their semantic characteristics. Vary those characteristics, and the explanation collapses. If the Dark Lord believed that lamb chops are made in Taiwan, we no longer have an explanation of why the world has suddenly become a New Zealand Free Zone.

Here's the point. In the last section, I sketched an argument that, if sound, shows that semantic properties aren't germane to the functional decomposition of the mind and the explanation of cognitive processes. For I claimed that cognitive processes are formal processes: keyed to the intrinsic features of mental states, not their semantic or representational features. For these consist in relations between us and the world to which mental questions have no unmediated access. Apparently, then, cognitive psychology leads to the view that the semantic properties central to intentional psychology *play no role* in the explanation of behaviour.[9]

Three responses are possible. One can simply conclude, as Stich is inclined to do, that intentional psychology is a false theory of the mind. Less drastically, one can urge that while semantic properties play no role in the explanation of *behaviour* there are *other* explanatory purposes for which it does play a role.[10] A third view is to suppose intentional psychology to be *incommensurable* with cognitive psychology. Rather than failing to integrate within a mature science of intelligence, intentional psychology is not a *candidate* for such integration.[11]

For my purposes, it doesn't matter which of these alternatives is accepted. For all show how folk psychology might be divorced from cognitive psychology. A collapse of folk psychology needn't threaten homuncular functionalism: the decomposition of the mind into a hierarchy of levels. Even if folk psychology fails, we will need a hierarchy of levels of psychological explanation. That fact leaves much open. It leaves open the contrast between Lycan-Dennett and Fodor, i.e. horizontal versus vertical functional organization. It leaves open a choice of primitives and their level. It leaves open whether those functions common sense identifies will survive as components within that hierarchy. But some form of hierarchial functionalism will be part of the story of how the mind works. It's much less clear that the computational element of our picture will survive in future theories of mentality.

3. COMPUTATIONALISM

The idea that cognition is computation faces some very nasty problems indeed. What follows is a preliminary sketch of some of these: a first rather than a last word.

Dan Dennett has characterised computationalism as the conjunction of the following three theories:

(1) Thinking is information processing.
(2) Information processing is computation (is symbol manipulation).
(3) The semantics of those symbols connect mind and world.

The defence of these three theses is the defense of 'High Church' Computationalism (HCC). The rest of this paper is on the problems of HCC. One of these has already been touched on: the problem of showing that there is explanatory point in the attribution of semantic properties to internal mental structures, granted that mental processes are keyed to the syntactic, not the semantic, properties of those structures. I will say nothing about that problem here, as I have wrestled with it elsewhere.[12] In this paper, I want to sketch two problems.

The first focuses on explanatory power. It's claimed that computational theories of mental processes have explanatory power, for we thereby are able to frame generalizations about the behaviour of systems that would otherwise escape us; this is one of the central themes of, for example, Pylyshyn.[13] I take this claim to subdivide into two subclaims. Firstly, the generalizations alluded to must be unavailable to those who refuse to indulge in computational vocabulary. They must be unstatable in structural, or 'merely' functional terms. The usual thought is that some systems are computationally alike, but physically heterogeneous, not only in detail, but to a considerable level of functional abstraction. The commonalities emerge only when we reach a computational level. The second claim is that these generalizations are generalizations about how the system works, not just what it does. Dennett, Newell, and Marr[14] all distinguish the specification of what the system does from any account of how it does it. Computational psychology is a theory of inner causes, not just a specification of performance capacities. I will suggest that its unobvious that these subclaims are jointly true.

The second worry has been aired by a number of critics. It's the idea that HCC is biologically implausible; computational models of cognition are unlikely to be true of messy, evolved creatures like us. So phrased, the worry is very vague. I will try out a few ways of giving it precise content, for I think there is something importantly right about this line of thought.

4. COMPUTATION AND EXPLANATION

Suppose that semantic properties do not have any place in a mature science of cognition. What would be left of the view that cognition is *symbol* processing? If the notion of representation is explanatorily redundant, and hence the brain processes no representations, then we scarcely have a computational theory of cognition. Rather we have "abstract neurophysiology – pure internal syntax with no hope of semantic interpretation."[15]

This issue might look verbal. But the screw can be tightened, especially if Fodor's modularity hypothesis is accepted. Fodor's model involves a tripartite division of cognitive mechanisms. At the skin there are transducers, which record the stimuli the world provides, and send on messages in neural code. Next come the modular 'input systems'. The raw data from the transducers are processed in these systems. The results are passed to the central processor, the site of the higher cognitive processes.

Input systems are conceived of as highly specialized; for example, there is not a single system for vision, but one for colour, one for shape, one for face recognition, and so on. These input systems or modules are 'domain specific', specializing in a particular kind of stimulus; they are 'innately specified', with structure not for the most part formed by a learning process. They are 'hard-wired', i.e. "associated with specific, localized and elaborately structured neural systems". They are 'autonomous', both in that they get no information from the central processes, and that central processes have "only limited (. . .) access to the mental representations that input systems compute".[16] That is, modules are 'informationally encapsulated'. They are 'mandatory'; there is no decision to operate, rather they are automatic, reflexlike.

What's the point of a computational theory of cognitive processes? Computational theories give us 'access to generalisations': a collection of systems that are physically or structurally heterogeneous nonetheless share properties that enable us to give unified explanations of their behaviour. These shared properties are computational properties. We can frame laws governing these systems; laws unstatable by those who fail to deploy computational notions.

The problem is simple. What justifies the salience of an intermediate level between a neurophysiological or structural account of (say) the perceptual modules and a semantic specification of the tasks they carry out? Modules are specialized, segregated and to a considerable extent, innately specified. Fodor expects both specialization and location in brainware corresponding to this functional of cognitive specialization. Hence it is hardly likely that there is not a rich body of neurophysiological and structural generalizations covering speech perception or pattern recognition. The systems, that is, are not so physically or structurally heterogeneous. Of course, there will be all sorts of local and minor variants across our cognitive peripherals. Just as there are, say, between different engines or bird's wings. But those variations don't prevent our discerning significant structural similarities in terms of which the workings of those locomotion systems can be explained. Just as a morphological description of wing structure doesn't have to be obsessively detailed, and can hence fit many wings, so too a neurophysiological description of our visual apparatus need not specify the location of every neurone; it can advert to the structural and functional commonalities across our different brains. The very strength of functionalism – the fact that it makes available functional vocabularies at a range of levels of abstraction – undercuts computationalism.

To the extent that our cognitive systems are modular, computational theories of cognition may well be redundant.[17] But so what? Not even Fodor thinks that *all* our cognitive processes are modular: our central processors are not. In any case, the modularity model is highly controversial anyway. Perhaps it's just wrong. Perhaps once we get past the transducers, we vary too much for any story but a computational one to work.

Abandonment of modularity yields only cold comfort to computationalism. Let's suppose that there is a good deal of flexibility, and hence plenty of individual variation, in the implementation of our central cognitive processes. Perhaps uhe contingencies of our different physical and psychological histories significantly affect the way our memory is organized, the pathways channelling information to and fro, our heuristics for problem solving, and the like. If so, then the wing-engine analogies of a few paragraphs back may fail. Our epistemic engines are too different for such similarities to be discerned. But why then suppose that these engines are *computationally* alike? Notice that once we move from the domain of processes that Fodor takes to be modular, capacity varies. Humans vary considerably in reading skills; mathematical and other problem solving abilities, competence in playing chess or recognizing music, and so on. In Dennett's terminology, qua central processor skills, our performance specifications are far from identical. This point shouldn't be pushed too hard: perhaps these differences don't reflect anything deep about the underlying cognitive processes, but rather result from a motley of interaction effects and other phenomena that are, from the perspective of psychological theory, just noise. After all, compared to the capacities of *other* species, human variation in – e.g. – the ability to learn chess is no doubt tiny.

But though the point about individual difference in capacity shouldn't be overstated, it need not be. For even if our capacities are identical, even if we were exactly as good as each other in the full range of such skills, it would by no means follow that those skills have the same computational implementation in each of us. Further, there is reason to doubt identical implementation: the reasons grounding our suspicion that there is individual variation in the structure of our central processor equally ground suspicion of computational difference.

Zenon Pylyshyn takes up this problem. When can we regard physically distinct systems as computationally alike? When are they computing the same function? The answer is in terms of functional architecture. It's *not* enough that the systems be weakly equivalent. It's not just that they are computing the same answer from the same data; they must be computing it *the same way*. That notion is spelt out in terms of architecture: two systems share an architecture if they share computational primitives. If the same steps are primitive for *A* and *B*, then *A* and *B* have the same architecture, and are computing in the same way, even if the physical realization of those primitive steps varies. But at this point a circularity looms. If we had a principled reason to regard the neural complexities of *A* and the different complexities of *B*, as both realizing a computational structure, and if we had a principled way of identifying the primitive elements of that structure, we could indeed apply (in principle) Pylyshyn's test, and conclude that *A* and *B* are computational twins if not physical twins. But that is *not* our situation. Rather, we wanted to know *why* we should discern computational structure in those neural complexities, and have been told: by doing so you can capture similarities that would otherwise escape you. This answer requires an *independent* ground for discerning structure in *A* and (possibly the same) structure in *B*; no such grounds have yet been provided.

In this section we have struggled with the question of whether computational models of cognition have explanatory power. They are typically defended as enabling us to discern (real, salient, explanatorily important) similarities amongst the physically heterogeneous systems. It's not obvious that they do. It's not obvious that there is a level between a functional but not computational account of the mind's structure and workings and the semantic specification of the performance of the system.

5. IS COMPUTATIONALISM BIOLOGICALLY PLAUSIBLE?

Thinkers of very different backgrounds have thought that there is something importantly wrong with computational psychology. Theorists as different as the Gibsonians, the Churchlands, Dennett and Hofstadter have suspected that computational models of cognition are biologically implausible.

The Gibsonians, for instance, claim that the data and methods grounding computational models of cognition are *ecologically invalid*. Irving Rock[18] is one important defender of the view that vision is importantly like thought. Rock's stimuli – and he is typical – are frequently two dimensional objects; when they are not, they are impoverished or eccentric in other ways. Little of the data appealed to consists of perceptions of three dimensional, natural objects. Instead of longish samples of bodies in motion, the stimulus objects are viewed for a very short time, or viewed by only one eye, or have unusual lighting or context, etc., etc. Fodor and Pylyshyn are brusque with this complaint.

> Gibson's criticism of these results is irrelevant once one accepts the condition (. . .) that an adequate theory must account for the effects of *all* perceptually effective stimulations. True we can alter the initial percept adding to the input (supplying context); but it remains to be explained how the original "ecologically invalid" percept was caused.[19]

This response isn't entirely convincing. True, we must *ultimately* explain all perceptual phenomena, including the nonstandard cases. But two responses are available to a sceptic. First, one wonders whether the processes that produce the various illusions, etc. that are the stuff of life for the cognitivist *are* one and the same as those operating in the wild, as it were.

Secondly, there is the question of idealization. Perception is a ferociously complex business; obviously not every aspect can be theoretically digested at the beginning. Some judgment about centrality must be made; to say the least, the phenomena investigated by cognitivists aren't obviously central: What's going on in the minds of the cognitivists is clear enough. By studying in the laboratory; by – e.g. – eliminating all depth cues they take themselves to be *isolating a single process* (say the detection of edges from discontinuities in luminosity) and hence making the study of perception manageable. But there is a tacit assumption here: namely the edge detection unit is isolated, so the very fact that no depth information is available from other sources has no impact on the edge-detection-from-luminosity machine. Well: that assumption *may* be right, but it's not *obviously* right.

The Gibsonian complaint is specific to theories of vision. Dennett, in a recent paper,[20] has tried to pose the intuition more generally. He quotes, with approval, Hofstadter's dictum that computational models of cognition are *too* efficient. Think of the functions calling subfunctions; routines deploying subroutines. All interacting efficiently; all coded – when they work at all – near optimally.[21] But optimal engineering – and that's at least what computational models aim at – is not Nature's way. For one thing, maximal efficiency requires detailed foreknowledge of the problem space. For another, it requires systems to be designed from the ground up. The processes of natural selection meet neither of these conditions. Stephen Jay Gould has emphasized that new biological adaptations are typically built from pre-existing structures that are modified through change in function. Adaptations are *not* designed from the ground up to fulfil their function. Further, natural selection has no foreknowledge: so in the evolution from an old structure or group of structures to a new one, all the intermediate forms must themselves be adaptive. Thus, typically, biological structures satisfice: that is, they are good enough to get by, but they are not optimal.[22]

It's hard to know how seriously we should be worried by this species of argument. For one thing, our intellectual capacities themselves are biologically mysterious. From the point of view of evolutionary theory, we seem grossly overendowed with intellectual skills. How did it help our ancestors to survive and reproduce at greater rates than their less gifted conspecifics when we consider skills like: the ability to see mate in three, the ability to prove theorems in number theory, the ability to recognize a tune, and many more? The answer is pretty obvious: these capacities were no help at all; somehow they must be free riders, consequences of *other* capacities which did help. But what capacities are these?

Dennett's worry is too vague to evaluate properly. I will consider two more concrete formulations of the same thought. Patricia Churchland[23] has argued that Fodor's own work shows that no computational theory of concept aquisition is possible.[24] Computational theories of cognition must assume that the concepts cognitive processes deploy are (in some sense to be explained) innate. The second more concrete formulation is also due to the Churchlands: they argue that computational processes are modeled on a linguistic, and hence parochial, model or representation. I will consider these ideas in turn.

5.1 Fodor and Churchland on Concept Acquisition

Consider what goes on when a concept is acquired. Perhaps someone is teaching you the concept 'wimp'. You will form a tentative hypothesis about the characteristics of wimphood as a result of your encounter, close or otherwise, with exhibited wimps. You will revise when your future uses are corrected, not otherwise. Eventually your hypothesis stabilizes.

In this example, this process might be relatively conscious. Mostly though, it will be completely *unconscious*. But even then, essentially the same process is going on. Moreover, this story is meant to be taken seriously: the hypothesis formation/confirmation story is *not* a metaphor. Concept learning involves the formulation of a hypothesis in some system of internal representation, a hypothesis then tested according to the canons of some built-in inductive logic.

As far as I can see, this is the only story *available* to a computational theory of learning. But it has a bizarre consequence. The only concepts that we can *acquire* are those whose extensions we can already *represent*. That is, we cannot acquire *new* concepts, at best only abbreviations of complexes of concepts we already have. Add in Fodor's (very plausible) claim that few of the words we use are definable, and we get a very strong nativist conclusion. Only obviously phrasal concepts like 'Left Big Toe' aren't innate, and they are built from concepts that are.

At first blush, this is a catastrophic consequence for computational psychology. Dennett invents the elegant expression 'cognitive wheel' for cognitive theories that aren't plausible for *creatures*. Fodor's view of concepts seems a cognitive wheel with bells on, for we are faced only with a choice between implausibles. Perhaps we have been paid in advance: nature has wired into us all just the concepts science and technology need many millenia before that could do us any good; many millenia, that is, before there could be any *selective advantage* in having those concepts available. Alternatively – and equally implausibly – one might suppose all the concepts we might possibly need are wired inside; we manifest 'BETA PARTICLE', 'TABLE' and 'FROG' rather than others because experience has 'activated' them. Finally, one might suppose that science and technology can advance only with respect to an *arbitrary* and *accidentally* prefigured conceptual stock.

In a fascinating paper, Fodor argues that his nativist view of concepts is much less implausible than it has so far been depicted. For to claim that concepts are innate is not to claim that they are present at birth, nor is it to claim that our experience of the world is irrelevant to our conceptual repertoire. Sexual development in humans is surely innate (we certainly don't *learn* to grow pubic hair, and the like) but it's not complete at birth, and it's affected by environmental factors, most obviously in its timing. Concepts aren't learnt, but they are *triggered*. That is, their development in us requires releasing experiences. The appeal to triggering helps.

(1) Since concepts are triggered by experience, Fodor can accommodate the fact that our conceptual repertoire is sensitive to our experience: different folks have different segments of their potential conceptual space activated in virtue of their different experience.

(2) Fodor's view is compatible with a naturalistic view of reference. Nativism would be utterly implausible if it were held that the *reference* of our concepts was fixed innately, for the Twin Earth stories and the numerous variations on that same theme seem to have shown that reference isn't fixed by *anything* internal to the thinker. Hence the referential properties of thought *couldn't* be fixed *independently* of the thinker's experience of the world. Allowing that innate concepts must be triggered by experience enables Fodor to grant this point without thereby granting that concepts are learnt: he thus leaves the door open to some form of a causal, naturalistic theory of reference.

(3) Much, much, more speculatively, Fodor suggests that we can explain conceptual growth and development by supposing that some triggers for concepts are, or include, or have as necessary preconditions, *other concepts*. 'QUARK' must be acquired after 'PARTICLE'. But not because 'QUARK' is defined in terms of 'PARTICLE' and other more basic concepts, but because among the triggering conditions for the quark-concept are other concepts.

I don't think that this salvage operation works. There is no way of rendering nativism both plausible and nontrivial. Furthermore, the notion of triggering that Fodor invokes turns out, on examination, not to be a brute causal biological notion but instead learning renamed. I suspect Fodor of employing a notion of learning that is, from the perspective of his own theory, forever inexplicable. Let's take those claims in turn.

(1) Nativist theories are always in danger of trivialisation. For a natural move is to deny the presence of innate concepts but to concede that there are, of course innate dispositions or capacities to aquire those concepts. This move trivialises nativism. Since Fodor insists on the essential role of triggering in the growth of concepts, how is his nativism to be distinguished from trivial nativism? Nontrivial nativism is (I think) committed to the view that pre-existing potential concepts are highly specific, and that triggering as a consequence plays a relatively minimal role in the activation of a concept. We can distinguish minimalist and maximalist conceptions of triggering. On the minimalist conception, triggering just tacks concepts onto reality: it's the triggering experiences that make the difference between one of our concepts being the 'WATER' concept rather than the 'TWIN WATER' concept. And that's about all triggering does. Alternatively, triggering may determine the internal functional role of the concept as well as its reference. It is in virtue of triggering experiences that some structure in Aristotle's brain realized the concept 'HAMMER' rather than 'BETA PARTICLE'. On such a view of the role of triggering, the pre-existing structures could not be richly and antecedently specified.

While the minimalist picture of triggering has some plausibility for sensory concepts and a few others for which there may be a specific evolutionary history (linguistic concepts? intentional and other folk psychological concepts?), *surely* it cannot be the right story for 'BETA PARTICLE' or 'XYLOPHONE'. But I think it *is* the minimalist notion that Fodor must have in mind. Not only is it needed to distinguish his position from a truistic one; the idea of a *specific* prestructuring of our potential set of concepts is implicated in his extensive defences of epistemic boundedness theories; the theory that various theories (including perhaps the true ones) are beyond our conceptual reach, however hard we try and whatever we experience.[25]

So construed, the addition of triggering to the computational theory of concept acquisition (an ersatz theory if there ever was one) renders it no more plausible than its unreconstructed ancestor. The same objection reapplies. For the tarted up version still requires the presence in our brain of highly specific structures anterior to their having any selective advantage. We have specialized and (in the naive sense) highly adaptive structures for which there is no evolutionary history. Surely that cannot be.

(2) Concepts aren't learned. But they *are* triggered. And thus nativism is saved from easy refutation. So saving it requires that triggering be a biological, 'brute causal' relation, *not* learning renamed. What then is it for a concept to be innate, i.e. triggered rather than learned. We could characterize this notion from two directions: by characterizing the notion of a trait under genetic control, *or* by characterizing the distinction between triggering and learning. These approaches converge. A trait is under genetic control if its appearance in the phenotype of the organism is insensitive to a wide variety of environmental conditions, namely all or nearly all of those normal to the organism. Thus, in humans, eye colour is under genetic control but body weight is not. The story about triggering goes much the same way: a cognitive structure is triggered *rather* than learnt if it is insensitive to variation in experience over a wide range. So (to take a classic case): new hatched ducklings imprint on just about anything that moves: 'if it moves, it's mother'. We take this to be an innate structure triggered or released by experience rather than a case of learning, for many very different experiences would have the same effect.

Triggering is thus distinguished from learning, in part, by the fact that a *wide range* of experience yields the same triggered element. Thus Chomsky is not abusing words in regarding universal grammar as innate. Though linguistic experience is required to trigger those structures, any linguistic experience will do.

The trouble is that the notion of triggering Fodor actually uses does not have this characteristic.

(a) The relation between triggering structure and releasing experience is very constrained. He writes:

> It is *not* part of the rationalist program to claim that your repertoire of primitive concepts is independent of the character of your experience. On the contrary, just as there are presumably triggering experiences that are nomologically necessary for the acquisition of sensory primitive concepts like RED, so there are presumably triggering experiences that are nomologically necessary for the acquisition of non-sensory primitive concepts like TRIANGLE.[26]

Concepts here aren't triggered by a wide range of experience. To the contrary: they are triggered by their instances.

(b) Furthermore, the process of triggering cannot be arational, brutely causal, for mere causal contact with appropriate instances is insufficient for concept acquisition. For one is thereby left with no account of: (i) empty concepts (why isn't 'PHLOGISTON' synonymous with 'OXYGEN'); (ii) which concept is triggered when the experience is ambiguous between many: my cat is a potential triggering of at least those concepts it instantiates, and these include: 'FEMALE CAT', 'CAT', 'FELINE', 'ANIMAL', 'MAMMAL' etc.; (iii) how concepts like 'NEUTRINO', 'WITCH', 'ROCKET', etc. can be acquired without experiencing their instances. While it's clear that causal contact with instances is central to concept acquisition, what else is involved is far from clear. But that else is unlikely to be nonrational: it's likely to be some set of beliefs and capacities involving that kind and its relation to other kinds.[27]
essentially inexplicable within computational psychology is bad news for that programme. First, because concept acquisition is a pretheoretic paradigm of a cognitive process. Second, because the view that we have a large stock of our concepts innately is, to put it mildly, implausible. Fodor attempts to defuse at least the second of these worries. Though unlearnt, these concepts are triggered. I claim his attempt fails. Specific pretriggered protoconcepts make no evolutionary sense. Further, his notion of triggering is a disguised, unexplained hence illicit cognitive notion. Computational psychology leaves us with no biologically plausible picture of concept acquisition.

5.2 The Infralinguistic Catastrophe

There is a second way the charge of biological implausibility has been made explicit. In a spirited defence of the relevance of bottom-up strategies for understanding the mind, Patricia Churchland has claimed that computational theories of cognition lead to an 'infralinguistic catastrophe'.[28] Her argument can be summarised as follows:

P1. Computational models of cognition depend on a linguistic model, for the representational structures they posit are languagelike.

P2. Linguistic models of representation are parochial. Language is evolutionarily and developmentally a latecomer.

C1. The representational or information processing systems of the brain are unlikely to be languagelike.
C2. So computational models of those processes are unlikely to be appropriate.

While none would pretend that this argument is apodictic, the premises do support the conclusion. But how well supported are the premises? Von Eckardt,[29] in her reply to the paper, takes issue mostly with the first premise. Once we see that a computational psychology is not mostly intentional psychology tidied up, P.1 just isn't true. Furthermore, some computational models of perception and imagery employ depictive rather than discursive representational systems. Even when we restrict ourselves to uncontroversially cognitive capacities, there are representational systems on the market that are not even linear.

At this point, there is an obvious reply to be considered. First, it's, to put it milely, controversial that depictive representations do have a place in the explanation of (say) imagery. Further, even if some depictive non-languagelike systems have a role to play in our cognitive processes, we cannot (without a much more radical rejection of folk theory than Von Eckardt and Kitcher[30] contemplate) withold a central place from languagelike systems. For depictive systems are insufficiently abstract (we can reason about something's colour without representing its shape) and insufficiently general 'we can reason about an object's mass or charge'. Moreover Fodor has argued, and continues to argue, that some pivotal systems of mental representation must possess the central semantic and syntactic properties of natural languages. According to Fodor, if we want to explain our ability to make decisions between competing options, or to solve a wide range of problems, or to plan behaviour, we must posit representational systems with the key semantic properties of natural languages: truth, reference, modality, tense. They must also possess the decisive syntactic feature: productivity from a finite base. For our capacity to represent in thought is unbounded in just the same way that our capacity to represent in language is.

True, this argument depends on accepting something like a folk theoretic description of our mental process. But neither Von Eckardt nor Kitcher want to defend a radical rejection of folk theory, so they are in no position to reject P1. That premise can be amended of course: some of our subsystems might use depictive or otherwise non-languagelike representational systems. But such a qualification leaves Churchland's argument essentially intact.

The assessment of the argument revolves around P2. That premise is very difficult to evaluate. There is certainly something to be said in its favour. There are different demands on a system used for internal purposes rather than for communication. There are also different constraints on the system: language must be vocalised and then accoustically decoded. Hence its linear structure. There are no such constraints on brain media. We can toss in Chomsky's insistence that the syntax of natural language is arbitrary, i.e. it's unpredictable from language's semantic and functional features.

Against this, two considerations must be weighed. Firstly, despite Chomsky's mulish silence on the questions, linguistic capacities must come from somewhere. Stephen Jay Gould, who emphasises the nonoptimality of evolved creatures, points out that this nonoptimality is typically a consequence of adaptations being old bits put to new purposes. From that perspective it wouldn't be too surprising if natural languages inherit some of the properties of the systems they evolve from. Especially, and this is the second consideration, if those inherited properties serve in their new function the same tasks as in their old. That seems likely: those inherited properties are very general semantic and syntactic properties, properties that make both language, and the ancestor system language developed out of, a productive and abstract system. All in all, I don't think Churchland's second charge succeeds.

The last sections of this paper have considered the thesis that computational models of cognition are biologically implausible. That's a very vague worry. When trying to make it more explicit, in giving it flesh, we have seen that it's not without foundation. Computational models of cognition are intriguing and suggestive. But there are good grounds for taking them to be, at best, tentative. The functional theory of mental organization, in one of its forms, must be a winner. Computational psychology could well be just wrong.

NOTES

1 J.A. Fodor (1983) *The Modularity of Mind*, Bradford/MIT.

2 Some founding texts for this theory are D.C. Dennett (1978) *Brainstorms*, Bradford, and W.G. Lycan (1981) 'Form, Function and Feel', *Journal of Philosophy* **LXXVIII**, 24–50.

[3] For a particularly good exposition of this point see D.C. Dennett (1978) 'Skinner Skinned', *Brainstorms*, Bradford.
[4] Lycan, in his 1981 (see *op. cit.*, note 1) discussion of 'qualia' objections, particularly emphasises this point.
[5] A classic text on vision is D. Marr (1982) *Vision*, Freeman. On imagery, see S.N. Kosslyn (1980) *Image and Mind*, Harvard. Perhaps needless to say, both Marr's and Kosslyn's work is highly controversial.
[6] And it does. We recognize faces on minimal information: caricatures, outline sketches, partly obscured photographs. Yet the process is idiosyncratic, too: we fail to recognise full portrait, but *inverted* photographs.
[7] D.C. Dennett (1986) 'The Logical Geograph of Computational Approaches: A View from the East Pole', *The Representation of Knowledge and Belief*, M. Brand and R.M. Harnish (eds.), University of Arizona Press.
[8] Patricia Churchland (1980) 'A Perspective on Mind-Brain Research', *Journal of Philosophy* **LXXVII**, pp. 185–207, and Paul Churchland (1981) 'Propositional Attitudes and Eliminative Materialism', *Journal of Philosophy* **LXXVIII**, pp. 147–170.
[9] See especially J.A. Fodor (1980) 'Methodological Solipsism as a Research Strategy in Cognitive Psychology', *The Behavioural and Brain Sciences* **III**, pp. 63–109.
[10] This view is defended in P. Godfrey-Smith and K. Sterelny (forthcoming) 'Animals and Individualism'.
[11] This view is defended in P. Pettit and J. McDowell (1986) *Subject, Thought and Context*, Oxford, and in D.C. Dennett (1981) 'Three Kinds of Intentional Psychology', *Reduction, Time and Reality*, R.A. Healy (ed.), Cambridge.
[12] K. Sterelny (1985) 'Critical Notice of S. Stich's "From Folk Psychology to Cognitive Science"', *Australasian Journal of Philosophy* **LXIII**, pp. 510–519, P. Godfrey-Smith and K. Sterelny, *op. cit.* (note 10), and M. Devitt and K. Sterelny (forthcoming, 'Fodor's Nativism'.
[13] Z. Pylyshyn (1984) *Computation and Cognition*, Bradford/MIT.
[14] Dennett calls this the level of performance specification, Newell calls it the knowledge level. David Marr rather confusingly calls it the computational level. The *real* computational level – if there is one – he calls the algorithmic level.
[15] See Dennet, *op. cit.* (note 11), p. 53.
[16] See *op. cit.* (note 1).
[17] I push this point in a forthcoming paper, 'Is Perception Computation?'.
[18] I. Rock (1983) *The Logic of Perception*, Bradford/MIT.
[19] J.A. Fodor and Z. Pylyshyn (1981) 'How Direct is Visual Perception: Some Reflections on Gibson's "Ecological Approach"', *Cognition* **IX**, p. 173.
[20] See *op. cit.* (note 7).
[21] I suspect that this point is exaggerated. It's commonplace to distinguish between theory relevant and theory nonrelevant features of a cognitive simulation. Naturally, the nonrelevant features are as efficient as possible. The relevant features need not be. There is no obvious sense in which – e.g. – Kosslyn's model of imagery use makes us hyper efficient: degradation of the image both over time and away from the point of focus are built into the theory relevant parts of the model.
[22] Stephen Jay Gould (1983) 'Perfection and Imperfection', *The Panda's Thumb*, Penguin.
[23] P.S. Churchland (1980) 'Language, Thought and Information Processing', *Nous* **XIV**,

pp. 147–170.

[24] See J.A. Fodor (1975) *The Language of Thought*, Harvester, and J.A. Fodor (1988) in this volume. Also see M. Devitt and K. Sterelny (1987) *Language and Reality*, Bradford/MIT for a more detailed response to this problem.

[25] See *op. cit.* (note 1).

[26] See J.A. Fodor (1981) *RePresentations*, Bradford/MIT, pp. 274–275.

[27] These complications are very great. See K. Sterelny (1983) 'Natural Kind Terms', *Pacific Philosophical Quarterly* **LXIV**, pp. 110–125, and M. Devitt and K. Sterelny, *op. cit.* (note 24), ch.5.

[28] See Patricia, *op. cit.* (note 8).

[29] Barbara Von Eckardt (1984) 'Cognitive Psychology and Principled Skepticism', *Journal of Philosophy* **LXXXI**, pp. 67–88.

[30] Patricia Kitcher (1984) 'In Defense of Intentional Psychology', *ibid.*, pp. 89–106.

Department of Philosophy
Victoria University of Wellington, New Zealand

BERNARD BEROFSKY

BELIEF AND RESPONSIBILITY

1. BELIEFS AS THEORETICAL STATES

1.1. Belief and Moral Agency

Few would deny that folk psychology embodies some essential components of our view of ourselves as persons. The specific element of personhood that interests me is moral agency, the feature that grounds judgments of moral responsibility. An agent is morally responsible for a state of affairs only if he is the sort of being who can generally direct his behavior towards (and away from) states of affairs. Thus, we need a psychological theory that permits effort directed to an end or action guided by an intention. This would certainly not be possible unless human beings have beliefs and behavior is a function of belief.

This position might be challenged by a proponent of an instrumentalist conception of responsibility. If one thinks that the justification of responsibility judgments and institutions is exhausted by their effectiveness in behavioral redirection, one will be indifferent to the psychological intermediary between judgment and change. An agent is responsible if he responds to such judgments, whether or not his beliefs are causally implicated.

I reject the instrumentalist conception, but will not argue the case here since I have done so elsewhere.[1]

Of course, to say that belief is necessary for moral responsibility is not to say that any conception of belief is a conception of a state required for responsibility. The fact that some elements of a psychological structure are designated 'beliefs' because they resemble the folk state in certain key ways does not automatically imply that the agent possessed of this structure satisfies the condition of responsibility in question. A functionalist account, for example, which identifies beliefs in terms of relations to inputs, outputs, and other mental states would be inadequate unless it had the resources to individuate belief-states finely enough. For the specific moral evaluation of a person depends on the content of the belief which guides action. Our attitude towards someone who is on the surface helping another varies, depending on the causally relevant description under which the agent acts, i.e., on the belief about the

Peter Slezak and W.R. Albury (eds.), Computers, Brains and Minds, 95–122

action which is a necessary condition of its performance. I may lend a helping hand to Fred, believing that he is a person in need, a friend, a person who will be useful to me, or a person I can psychologically destroy by rendering him dependent on me. My moral worth hinges on counterfactual inquiries as to which of these beliefs was necessary in the circumstances for my action.[2]

Thus, those radical versions of eliminative materialism which anticipate a displacement of psychological theory so complete that we must regard the intentional concepts of folk psychology as being then discarded are committed to the disappearance of the moral evaluations of human beings resting upon those remnants.

1.2. Concepts of Belief

The contemplation of this stage may be rendered more palatable by the conviction that the rejection of beliefs as useful theoretical concepts does not automatically require the elimination of belief-talk as outmoded. But can such an extreme divergence between science and common sense persist indefinitely? Would we not anticipate an eventual erosion of the commonsense idioms now entrenched so deeply as to make it virtually impossible to imagine what it would be like in such a world?

Commonsense idioms coexist comfortably with scientific ones when scientific theories are applied to phenomena individuated by concepts in ordinary language. To explain why grape stains are difficult to remove or why rhubarb leaves are poisonous is not to place these generalizations in jeopardy.

This argument, however, would appear not to be available in the current context to the optimist, i.e., one who envisions a prevailing spirit of cooperation between science and common sense. For the commonsense generalizations which survive theoretical explanation link terms embedded in ordinary, pre-theoretical language. (Of course, this is a relative matter. Botanical discoveries might cast doubt on the viability of the category 'rhubarb plant'; but it will take a lot more to achieve this end than the discoveries which explain the poisonous character of the leaves of these plants.) Beliefs, on the other hand, are supposed to be the very states which explain behavior, i.e., the states of mind which, given their complex relation to others, intervene between stimulus and response and tell us why that stimulus led to that response.

They are, therefore, *competitive* with alternative theories of behavior in a way the fact that this is a rhubarb leaf does not compete with alternative accounts of the poisoning. The well-confirmed empirical generalization linking rhubarb leaves and poisoning contrasts with a theory of behavior in which belief states are invoked as *theoretical* states of the system rather than noncontroversial input or stimulus conditions.

But is this clearly so? The thoughts which naturally induce philosophers to construe belief as an internal theoretical state, linked indirectly to belief-expressions, have a flip side. The more internal beliefs get, the less belief-like they become. Cognitive theories tend, for example, to treat them exclusively in terms of content. What gets processed or stored in the mechanism is information, whether it is treated sententially or in terms of more remote units, such as lists, problem spaces, or bytes. We look kindly on computer simulations of thought when we do not expect simulations of the other aspect of belief, mode. Belief is a different mode from desire, although they may be identical in content, e.g., that I will be alive tomorrow. Many mental models concentrate on content, not mode, as if storage in the human mind is a phenomenon like storage in a memory bank. The question is whether or not it is there, i.e., the sentence or whatever, not whether or not it is affirmed (or barely believed or under evaluation or . . ., where a variety of cognitive modes may be given).[3]

Perhaps modal distinctions, then, must be defined on the behavioral level, but in a way which permits us to retain the idea that beliefs are internal states. An internal sentence is a belief when, typically, it explains disposition to *assent* behavior, a desire when it typically explains disposition to *seek* behavior, etc. Of course each such state can also explain a variety of types of behavior in addition to the type which is mode-determining for that group.

What is a disposition to assent? As Dennett has said, it is not a disposition to utter or say. I may be disposed to say many things I do not believe. He notes, "sometimes we salt away a sentence because we like the sound of it, or because we will later be rewarded for producing it on demand, or just because it has a sort of staying power in our imagination".[4] The key question then becomes the difference between saying and asserting (or assenting).

If we define assent as a feature of conscious propositional episodes ('thinkings-that-*p*') or verbal utterances, i.e., a feature available to its owner in a privileged way, belief as an internal state explaining behavior, including assent, will be different from – perhaps structurally very different from – one of its effects, assent.

Perhaps, as some have said, we really need two concepts, belief (internal) and judgment (assent).[5] In addition to the reasons cited above, there is the oft-cited difference between the either-or character of judgment and the various notions of degrees of belief, such as Bayesian ones, which permit important distinctions among our epistemic relations to sentences, most significantly our assignments of subjective probability. "The effort of retrieval is often an effort to formulate a sentence that is an approximation of a belief, and we are often distressed by the hard edge of determinacy our verbal output substitutes for the fuzziness of our convictions."[6]

And if our judgments can be known by us to be sincere, to be assents, the inference to belief is nonetheless corrigible. The failure of dispositional accounts of belief is based in part on the complexities explaining this fact. My other behavioral dispositions, for example, may justify the conclusion that I do not believe that to which I assent. I suffer from self-deception. Although the belief that *p* usually explains assent to *p*, in unusual cases there is divergence.

So a case for the distinction is strong. It might also explain our ambivalence about ascribing beliefs to animals in terms of the fact that they have nonlinguistic beliefs, but make no judgments.

If the resurgence of 'believing' on the phenomenal or behavioral level permits those who cannot bear the total abandonment of this concept to suffer the distortion beyond recognition of belief as an explanatory theoretical concept, they must at least be sure that what they have salvaged is not utterly ersatz. What precisely is the feature of judgments that plays the role of believing?

Assent does appear to have a feature we are looking for, privileged access, a feature beliefs lack. For if assent or acceptance is an action which represents some sort of commitment to the truth of the proposition accepted, we know we have performed it, even if we have never performed it for the vast majority of (internal) beliefs, e.g., the infinitely large class of bizarre logical consequences of our mundane beliefs.

But assent cannot be the act of commitment for I can commit myself publicly to propositions I clearly disbelieve.

However assent is ultimately characterized, the concept of belief that is implied in intentional or effortful action *is* an explanatory one – it construes belief as causally related to behavior. If, therefore, belief is rejected as an explanatory notion, what we will be left with, felt confidence or assent, will be too weak to sustain our current conception of a morally responsible agent. As Dennett notes, the states arising from acts of assent are a poor guide to prediction of behavior unless they coincide with (internal) belief.[7]

We should also remind ourselves that, as rational beings and language users, the concept of (internal) belief is fundamental in relation to assent. Most of our beliefs and the principles of rationality to which we subscribe are implicit, and we praise and blame people for acting on them. We even praise and blame people sometimes for acting on beliefs with respect to which they are victims of self-deception. And, as Cherniak has noted, if assent is a sufficient condition of (some sort of) belief, the latter state cannot play the role it does in attributions of minimal rationality. For attribution of the belief that *p* would not automatically implicate *any* logical consequences of *p* or of *p* in conjunction with other intentional states known to be possessed by the agent.[8] In other words, we cannot ascribe beliefs singly if we think people are minimally rational; but the assent interpretation permits that.

Thus, the concept of belief implicated in moral responsibility is the concept of a theoretical state and, therefore, the retention of the notion of moral agency depends in part on developments in psychology.

2. THE A PRIORI CASE AGAINST BELIEF

2.1. Beliefs as Individual Psychological States

The threat to belief which has of late engaged the attention of many philosophers arises independently of specific developments in cognitive psychology. As McGinn has noted, the concept of belief is strained by its double-duty use.[9] When we communicate our beliefs, we think of them as possessed of truth-conditions, and when we invoke them to explain behavior, we treat them as intrinsic states of the agent, e.g., representations whose semantic properties, as relations to the world beyond the agent, are causally irrelevant to behavior.

> These components and the concerns they reflect are distinct and independent – total content supervenes on both taken together. We get different and potentially conflicting standards of individuation – and hence different conceptions of what a belief essentially is – according as we concentrate on one or other component of content.[10]

Fodor, Putnam, Burge, Stich, Kaplan, and others have advanced considerations on behalf of this conclusion. The basic idea is that ascription of content to a belief-utterance depends to a significant extent on facts independent of either the internal representation or the physical state of the person. Putnam's Twin-Earth cases show that persons can be intrinsically identical, physically and psychologically, yet their syntactically identical belief-utterances (*Water is wet*) have different meanings.[11] Burge's arthritis case[12] and Stich's endive-chicory case[13] reveal the social dimension of meaning, i.e., the necessity to interpret an utterance in terms of features of the linguistic environment of the speaker which may have had no causal impact on the person. Again, we are provided with illustrations of identical speakers uttering identical sentences, yet meaning different things by those sentences just in virtue of their membership in different communities.

For Fodor, the moral of the above is that semantic features of representations can play no role in a science of behavior and cognitive psychology must, therefore, restrict itself to formal operations.[14] Stich adopts a principle, autonomy, similar to Fodor's methodological solipsism, according to which psychologists are restricted to states and processes which supervene on the internal physical state of the organism. Historical and environmental facts relevant to content ascription must be ignored unless they make a physical difference to the organism.

The project naturally generated by the above reflections is the bifurcation of belief into its two components. "The explanatory force of the content ascription attaches only to the contribution the words in the content clause make in their capacity as specifiers of internal representations; their referential properties play no explantory role."[15] What may we expect is left over when we remove content or semantic representation? A natural answer, Stich's, is purely syntactic objects. Another, McGinn's, is cognitive role. This idea is explained in terms of Field's probabilistic semantics in which a sentence's cognitive role (Field uses 'conceptual role') is determined by the subjective conditional probability-function on that sentence.[16] The intuition is that these assignments are based not on the state of the world, but rather on the speaker's conception of that state, for example, on the description *he* assigns to the names he uses. A third idea, Dennett's, is notional attitude.[17] Rejecting a purely syntactic account such as Stich's, Dennett proposes that we retain semantic interpretation, but fix it by environments which may or may not be *real*. We can characterize the organismic contribution to the determination of propositional attitudes in terms of the individual's 'world', regardless of the extent of its coincidence with the actual world. Others, like Burge, are skeptical of the strategy dictated by methodological solipsism and prefer to rethink its individualistic orientation.[18] It is preferable, he suggests, to reconstruct our conception of psychological explanation along social lines rather than to define the bifurcation which would, if it could be produced, permit us to retain the individualistic orientation some see as virtually a conceptual requirement for psychology.

What is the bearing of all this on our concern? The skeptical argument we must address is that beliefs cannot guide behavior (a condition of moral responsibility) because beliefs are essentially individuated by content and are, therefore, not individual psychological or physical states.

It is clear that if Robinson acts the way he does because he believes that water is wet or that aspirin relieves arthritis and is instantaneously and secretly whisked to Twin-Earth where, not only is water *XYZ* rather than H_2O, and the speakers there use 'arthritis' to refer not only to what we do by it, but also to aches in the thigh, the change in belief-content necessitated by the switch will not change Robinson's behavior suitably redescribed.[19] One unexciting consequence of this is that behavior guided by a belief would persist if the belief were replaced by a different one, so long as the replacement belief is related to the original in the right way. This conclusion does not require us to give up the position that beliefs are necessary conditions of actions. In the real world, Robinson would act differently if he were to surrender the belief that water is wet, because that belief will *not* be replaced by an equivalent one. The fire would not have ensued if the match had not been struck because there was no other combustible material present. And even if a fire would have occurred in some other way, a way preempted by the match striking, we still pick out the match striking as the cause. Similarly, Robinson is watering his plants because he believes that $water_{H_2O}$ is wet, even if God dictates that the only way Robinson can surrender this belief is by replacing it with the belief that $water_{XYZ}$ is wet.

Notice also that we can be confident that the striking of the match caused the fire even if we cannot produce a precise definition of the replacement class, i.e., the class of event-types tokens of whose members invariably lead to fire in the presence of oxygen and 'sufficient' heat.

This simple point may be buttressed by skepticism pertaining to the whole project of characterizing organismic contribution in order to insure the 'intrinsic' nature of belief. One of the more obvious lessons of relativity theory is that fundamental physical properties, e.g., the length of a moving object, are not intrinsic, but are rather relative to a frame of reference which is stationary relative to the moving object. It is not that changes in the frame of reference *cause* a change in the length of the object. That would be a reversion to the idea of intrinsic length. There is rather an additional term, the frame of reference, in the specification of length such that changes in that term constitute different specifications. This fact does not prevent us from thinking of length as a feature of the moving train nor from the formulation of laws about magnitudes relativistically characterized which may apply to causal processes. Since physics is everyone's model of propriety, why can't we analogously say that beliefs are relative to physical and social environments such that relevant environmental changes require us to attribute different beliefs to a person? The beliefs would be the person's beliefs; they can, like length, be causally related to other things; relativization, we must recall, is not the incoherent supposition that causally irrelevant environmental changes must somehow affect (?) the person.

The response to the above which highlights the motivation of a principle such as Stich's autonomy principle is that beliefs, so construed, could never be 'reduced' to or supervenient on states of the brain. The real issue is not between intrinsic or relational construals, but rather between conceptions of belief which do or do not render it supervenient on neurological states. After all, beliefs, like anything else, can be picked out relationally. For purposes of explanation, however, there ought to be a way of picking them out such that on these descriptions, a knowledge of the neurological state of the person suffices for determining his belief state.

But if materialism says that beliefs *are* neurological states, why not pick them out (someday) in terms of neurological properties? On Davidson's anomalous monism, in fact, those properties are the ones under which the states are causally efficacious.[20]

Although anomalous monism strikes me as an incipient version of eliminative materialism, even if we concede that it permits the existence of beliefs, it certainly does not permit moral agency. If I act in the way that I do 'in' the belief that I am aiding Wilton, I am praiseworthy *only* if I would have acted differently had I believed that that action would have destroyed Wilton. But anomalous monism does not guarantee such counterfactuals because it supposes no nomic connections between the causally efficacious neurological properties and morally relevant psychological properties.

Let us then see where we are led when we seek to extract 'speaker's meaning' from the truth-theoretic component of meaning in order to characterize sets of replacement classes, i.e., classes of equivalent beliefs.

2.2. Theories of Cognitive Role

On Stich's syntactic theory of the mind, Robinson, the fellow whisked to Twin-Earth, is an easy case. Since the switch changes neither the sentence-type he 'believes' nor the behavior it partially explains, no changes in his psychological state have taken place. For Field, analogously, the conceptual role of *Water is wet* will not have changed for Robinson because the sentence prior to the switch is equipollent to the sentence after the switch, i.e., he continues to assign the same (subjective) probability values to the sentence conditionally on other sentences. (How can these epistemic facts change in light of a change – H_2O to *XYZ* – having no effect on Robinson?)

If equipollence defines the replacement class, we can give more precise content to the idea that a belief guides action in that the action would be different were the agent to surrender the belief or any other equipollent to it. We might even imagine that psychological theory will identify cognitive role as the property of beliefs in virtue of which they have causal efficacy.

There is another advantage to the endorsement of an approach like Field's. Since content individuation of beliefs creates psychologically irrelevant distinctions, we seek sets of beliefs (equipollent sentences) such that all the members of one set have the same total causal potential.

It turns out that we must adopt a strategy similar to this one for independent reasons. The discussion so far has supposed that beliefs are individuated by reference to content and the content specifiable by a sentence. If beliefs (in part) explain behavior, then, organisms must enter sentential states, i.e., certain neurological events or states must be identified in terms of or related to sentences, so that reference to those states will constitute a (partial) psychological explanation. We may call the view that there are these 'mental sentences' and that they play a crucial causal or explanatory role 'sententialism'.

An argument that the existence of beliefs does not presuppose sententialism is provided by the infinity of beliefs. Even if beliefs are not generally closed under logical consequence, I may be rational enough to believe every or almost every proposition – and there are an infinite number of them – following by the rule of addition from some belief of mine. But there is not an infinity of neurological states to serve as vehicles for the representation of each belief. So, the argument concludes, I must 'store' beliefs in a way different from that suggested by sententialism.

The argument would appear to prove too much. If being 28°C is a state of some liquid, why wouldn't this state 'generate' an infinite number of states, e.g., being 28°C or being 29°C, being 28°C or being 30°C, etc.? Since we do characterize states with varying degrees of specificity, it is not plausible in rebuttal to deny that the disjunctions are not real states of the liquid.

It is, however, plausible to think that the precise temperature state generates all these *nondistinct* states in a way an atomic belief fails to generate nondistinct disjunctive beliefs. We are not, therefore, perplexed about the prospect of an infinite number of states of the liquid in a way we might be perplexed about an infinite number of beliefs. The reason is the very failure of closure in the belief case. It is *not* utterly trivial to learn that I accept the logical consequence of some belief. Thus, belief in the conclusion of a valid argument is a 'separate' fact from belief in the premise, and we must again confront the dilemma of relating an infinite set of beliefs to a finite set of neurological states.

A more promising model is generated by reflection on the fact, noted by Descartes,[21] that a single state of a physical system can be associated with an infinite number of dispositions. To use Descartes' own example, a piece of wax can flex in an infinite number of ways, even if its flexibility is identified with an underlying structural trait. The analogy breaks down, however, if we individuate beliefs by content because the underlying neurological state would not be the belief. To make the fine-grained distinctions required for content individuation, we would have to identify beliefs with dispositions. For just as a piece of wax can flex an infinite number of ways, so in a sense am I disposed to assent to an infinite number of sentences. A neurological state would perhaps explain why a person has a *set* of beliefs; but since there is a finite set of distinct neurological states, they cannot literally be identified with beliefs.

Thus, again, content individuation is too fine. Suppose Webster, an intelligent, educated, adult American has just died. He never contemplated the sentence *Ronald Reagan is not an Indian princess* (*R*) nor the distinct sentence *Ronald Reagan is not a princess* (*P*). It is very probably true that he believed both. (It is not certain, however, for he may, for all we know, have been suffering from a localized mental aberration which would have manifested itself in a narrow set of circumstances none of which ever obtained.) There was not, however, an intrinsic state of Webster we can identify with *R* such that it is plausible to assign a distinct state to *P*.

Now *R* may not be equipollent to *P* for Webster. There may be a wild set of circumstances under which he would assign only a fairly high probability to *P*, yet regard *R* as pretty certain – he is quite sure Reagan is not Indian, even if he has some reason to believe 'he' is a princess.

Now if Webster believed both *P* and *R*, surely, as the autonomy principle says, his total physical state determined these facts. And if there are too many beliefs to identify with physical states, we must, as we said earlier, think of physical states as determining a set of beliefs. The fact that neurological states are not beliefs together with the fact that the mapping of beliefs to states is many-one does not entail rejection of the autonomy principle. A complete physical description determines a unique set of beliefs. (Beliefs on this account are closer to properties of states, e.g., believing *P* and certain logical consequences of *P*, perhaps those logical consequences of *P* Webster's logical acumen, factual knowledge, and psychological makeup 'permit' him to accept.) Thus the set of beliefs determined by some physical configuration includes all beliefs equipollent to any in the set *plus others* (as the example of *P* and *R* makes clear).

How do the above considerations affect the requirement of responsibility that beliefs guide action? We had earlier contemplated the possibility that a belief is causally efficacious via the property of cognitive role it shares with all equipollent beliefs. We are now suggesting that a (probably highly complex) neurological state determines a set of beliefs, but cannot literally go on to say that beliefs determine behavior. For the ontological status of beliefs, since they are not neurological states, is now unclear.

These results accord, for very different reasons, with those of McGinn and Stich, who contend that the underlying explanatory states picked out by belief-sentences are not beliefs. But if belief-sentences are true and stand in causal implication relations with behavior-sentences, can we then hedge on the claim that moral responsibility requires *beliefs*?

This strategy is plagued, however, by the fear that we are retaining the truth of 'My belief that *p* is ——' in spite of the fact that the subject term does not refer to a belief only because we are not quite psychologically (?) ready to capitulate completely to a radical eliminativist position. We are not as comfortable here as in an intersubjective context in which referential success is achieved in spite of misdescription. I get to tell you that the man standing in the corner is angry because you know to whom I am referring when I say *The man drinking the martini is angry* for, although the man in the corner only looks as if he is drinking a martini, I am referring to him.

I think that moral responsibility is salvaged if we acknowledge that behavior is a function of the belief-determining property of neurological states even if those states are not beliefs. We saw earlier that belief-guidance is compatible with behavioral invariance under transformation within a set of (equipollent) beliefs. Now just as behavior can be a function of several intentional states, so may the set of all beliefs causally relevant to a bit of behavior include several beliefs not equipollent to the ones in the original set. For example, if I stare at Fred because I know that it will embarrass him, I may be blameworthy for this behavior. I have, but never consciously think of, the belief that embarrassing people will not cause me to be struck by lightning. Were I to believe otherwise, I would not embarrass Fred. I am still morally responsible for staring because a causally relevant property of my physical state is mapped to a set of beliefs (including the belief that embarrassing people does not cause one to be struck by lightning) which includes the belief that staring will embarrass Fred.[22]

Although there are all sorts of questions about these properties, the basic point I am making is similar to Stich's: "(. . .) generalizations specifying causal relations between stimuli and neurological states will identify the neurological states not by adverting to their essential neurological types but, rather, by adverting to the syntactic objects to which the neurological types are mapped".[23] Many object to Stich's choice of syntactic objects; but cognitive science would not exist unless it had a viable substitute for the identification of neurological states via neurological properties.

2.3. Reconciliation of Competing Accounts of Cognitive Role

If Stich insists that the causal role of beliefs is grounded in the formal relations between the syntactic objects to which psychological states are mapped, rather than the cognitive role of those objects, even if they have such roles, why may we not, in a conciliatory spirit, suggest that the causal efficacy of Robinson's current neurological state is based upon its status as a token of the very sentence specifying the cognitive role of a 'believed' sentence? That is, if Robinson speaks language L, there will be a set of probability-functions defined on L each of which maps pairs of sentences into real numbers. Some of these functions obey the axioms of probability and certain conditions of reasonableness which are not important here to specify. Suppose we discover – and there are several problems pertaining to the conditions of this discovery we shall discuss below – that one of these reasonable (subjective) conditional probability-functions (call it P_r) is Robinson's at t. Suppose also that E is 'There is now an elephant in the field' and that the restriction of P_r to the unit set of E, i.e., that portion supplying probabilities to all the pairs whose initial member is E, is a function we shall call P_{re}. The conceptual role of E for Robinson at t just *is* P_{re}. For example, given that Robinson believes that F ('My current visual field includes a large, gray animal with big, floppy ears and a long trunk') and that P_{re} has $P(E/F) = 0.9$, Robinson begins to seek peanuts. So we have a sketch of a causal explanation precisely along the lines of the syntactic theory, i.e., as one based on formal features of the objects to which the neurological states are mapped. The details will fill in the links so that semantic information will become unnecessary. For example, we will have facts like 'Robinson believes that elephants like to eat peanuts', 'Robinson wants to find what elephants like to eat', 'Whenever Robinson wants something, he is likely to begin to seek it', etc.

The conceptual role of E plays a crucial role in the explanation. P_{re} determines that certain beliefs will be sufficiently probable so as to be able to give rise to relevant behavior.

If this is right, then the accounts of Stich and Field are complementary rather than competing. Stich's account can incorporate Field's in the above way and Field himself has said that there is a belief relation whose second term is an internal token and that "the sentences in the system of internal representation might as well be meaningless as far as the psychology is concerned".[24]

Before we wax too ecstatic about this harmony, we must identify problems Stich can point to on his behalf. Field himself notes that the determination of a conditional probability-function is not only a highly complex affair, but one which represents an idealization. The verbal and nonverbal data constituting the evidence, no matter how massive, would not permit a unique assignment without a variety of conventions, including assumptions of rationality. Conventions are crucial even if we decide to permit indeterminate probabilities or substitute upper and lower probabilities for completely determinate assignments.

Idealizations, the need for conventions in empirical inquiry, the gap between theory and evidence are all familiar features of scientific investigations into both the natural and human realm. There is one element, however, which Davidson and Putnam regard as sufficiently special to preclude the satisfaction of Stich's autonomy principle by beliefs in general or beliefs assigning conditional probability-functions in particular. Beliefs of the latter sort in fact display this feature most dramatically. For, as suggested above, they cannot be ascribed without an extensive range of interpretations of other verbal and nonverbal behaviors, incorporating meaning assignments and belief attributions, which permit even the most irrational person to be counted as a member of the community of believers. The situation is *radically* different from one in which many empirical assumptions are required before an inner state can be postulated. The assumptions Davidson and Putnam have in mind are not *a posteriori* at all, but rather posits permitting us to construe the entities with which we interact as rational beings.

For Putnam the elaboration of these conventions is not a part of psychology at all.[25] There are many ways of doing it. We were naive to think that scientific psychology could in principle replace the casual, occasionally insightful observations of the humanist, or that the interpretations of humans would become more 'objective' than the interpretations of texts.

If these two are right, a subjective conditional probability-function, embodying in its application the community's constitutive ideals of rationality, fails miserably to pick out an intrinsic characteristic of the believer, one insulated from causally independent factors such as the community of which the believer is a part.

2.4. The Social Nature Of Belief

An alternative to the syntactic theory would then require a rejection of the Davidson-Putnam line, or at least that aspect of it which threatens the idea that a belief is an internal state of the believer. The threat derives from the view that belief-attribution represents in part a feature of the attributor, based on an unavoidable commitment on his part to interpret behavioral data as products of a rational agent (even if he is acting *very* irrationally). Interpretation here is not empirical in character, but represents rather a kind of social Kantianism, one intrinsic to a community *a priori* committed to treat its members as persons or agents.

It goes without saying that an ascription of rationality runs very deep and is presupposed in all our dealings with other members of our community. But, to seek an analogy in the physical domain, so is the belief in gravity. Does this mean that there are postulates of physical theory that deserve to be separated out as *a priori*?

As to Davidson's contrast argument, to wit, that I do not understand an ascription of irrationality unless it is advanced against a background assumption of basic rationality, it does not follow that I must assume that *that* irrational agent is basically rational. I can imagine degrees of irrationality leading to the loss of personhood. We tend to be sympathetic with Davidson's argument because we realize that even the most extreme sorts of mental breakdown leave systems of verbal and nonverbal patterns of rationality reasonably intact. This is to be expected considering that, as a species, we are hard wired for linguistic competence. But it is easy to *imagine* a breakdown so severe that we can place no plausible interpretation on behavior. And if I can imagine this for one 'person', I can imagine it for my 'community' – a real nightmare.[26]

To be sure, I am imagining this from the perspective of my rationality. And I could not discover this or anything else were I not rational and were there not a host of other conditions present – but one of these is not the rationality of the object of investigation.

But even if we can deny that the being before us is in any way rational, is not the constitutive ideal implicated as soon as we impute some belief?

First of all, an adequate account of cognitive role permits holism without surrendering the idea that beliefs are intrinsic. The cognitive role of a sentence is not sufficient for behavior. Any behavioral prediction depends as well on information concerning a variety of beliefs.

More importantly, the beliefs which are implicated, directly or indirectly, by a given belief will be related in appropriate ways by the probability assignments. (In the simplest case, if Jones believes Q and R is a logical consequence of Q, then for any S, $P(Q/S) < P(R/S)$.)

Second, suppose we have finally found a way to link individual belief states with individual brain states. Suppose then we discover that the very beliefs presupposed or implicated when a belief state is attributed are associated with brain states which are themselves nomically related. So holism is reflected in the physical domain by nomic connections. I think materialists would be rightly more impressed with these results than with the objection that the nomic relation in the physical sphere is weaker than the conceptual relation in the psychological sphere in that the sense in which one belief presupposes another is stronger than the sense in which one state nomically implies another. If these results confirm physicalism, holism is compatible with the material and, therefore, intrinsic character of beliefs.

More significant than holism is what I shall call 'communality'. When we impute a belief, Davidson notes, we presuppose not just a network of beliefs, but that we, the interpreters, share the bulk of these beliefs.[27] We cannot place an adequate interpretation on an utterance unless we share the vast majority of beliefs implicit in the concepts and logical operations we invoke in the interpretive act.[28]

Communality would appear to threaten the individual nature of belief by positing as necessary to a belief the extra-individual fact of sharing. Moreover, equipollence, which earlier permitted us to nullify the changes on belief content wrought by social and physical environmental variations, is of no use here. I should like to argue that this threat can be dismissed in a way which will demystify this most recent effort to erect an impassable gulf between the *Naturwissenschaften* and the *Geisteswissenschaften*.

It is commonplace to note that the application of a concept which is a part of a scientific theory carries with it a variety of lawlike assumptions. To classify behavior as mammalian is to place this behavior in a context which presupposes that the object is an animal who has evolved in a certain way and shares a variety of structural and behavioral traits with other animals similarly classified. It would take just as enormous a breakdown of familiar, background assumptions for me to withdraw the conclusion that I am observing mammalian behavior when I see what I would normally describe as a chimpanzee suckling its young as it would to withdraw the conclusion that Jones has expressed the belief that it is raining when he utters the words *It is raining*.

There are differences, to be sure. An imputation of rationality, not mammalhood, brings with it an infinite number of other states. But this is an issue separate from communality and already addressed in a way which preserved the intrinsic character of believing at the cost of some ontological reshuffling. The most interesting difference, which is responsible for the inscrutability of communality, is that the interpreter is himself a member of the very class he is imputing to his subject, i.e., the class of believers. That is precisely the reason he must have so much in common with his subject.

Of course, I am a mammal too. But the imputation of mammalian status is not *ipso facto* a mammalian act (although only mammals do it). Imputation of anything *is*, however, *ipso facto* an act of belief. So it is a part of the logic of investigation that the study of belief presupposes a social context in which the sharing by scientist and subject of a variety of intentionalistic assumptions must be taken for granted. This no more justifies the cryptic conclusion that beliefs are social[29] than the fact that all mammals share a great deal in common justifies the conclusion that mammalian states are social. Communality derives from the sharing of intrinsic states implicit in membership in the same class.

But if two beings are believers, why must they share the bulk of their *specific* beliefs? This admittedly more dramatic similarity derives from biological and psychological similarities, possession of a common language (or universal grammar), similar individual histories and, possibly, social relationships. The fact that vast similarities are presupposed in belief attribution is not an argument that the similar states are not intrinsic ones.

The rationality posit is, to be sure, rooted in biology *and* practical necessity. We will not be easily convinced that we are in the nightmare earlier described. So the posit is deeply entrenched. But as Putnam says when reconstructing Davidson's argument, there may be *no* level at which we are as rational as we often think we are. This is true and reflects not the bifurcation of two enterprises, functionalist psychology and interpretation theory, but rather the naiveté of available theories in functionalist psychology. Many theories make us far more rational than we are. There are reasons for this, e.g., it is methodologically simpler to formulate rational decision theory *a priori*, then adjust for the realities. But we must remember to adjust![30]

With respect to our concerns, these adjustments (e.g., the rejection of simple transitivity of preference) do not necessarily imply the abandonment of belief. This is a bald claim, one I am not making good on here. But it seems to me that one can go a long way in acknowledging the irrationality of people without surrendering the belief in belief.[31]

I will not decide here between the syntactic theory and one based on cognitive role. Although the latter is in better shape than some would have us believe, new accounts of organismic contribution may appear which merit our attention. I see no *a priori* barriers to the development of a theory of intentionality in which beliefs are construed as individual psychological attributes which enter into explanation of behavior.

3. COGNITIVE PSYCHOLOGY

3.1. Weak Sententialism

It is clear to all that cognitive psychology is too young to permit reliable predictions as to the character of the representational structures which will eventually emerge as the dominant ones. Although it may, therefore, be premature to announce the death of sententialism, the doctrine that sentences are the basic representational vehicle, it is certainly premature to conclude that a far weaker doctrine, what I will call 'weak sententialism' is moribund. Weak sententialism demands only that sentential structures play some significant role in our mental architecture, but does not preclude the inclusion of other representational forms. To explain action by a mental machinery which invokes sentences does not preclude the incorporation of the rest of the cognitive psychologist's arsenal. If frames, arrays, lists, semantic nets, and images are important, why not sentences as well?

The importance of sentences in this context is grounded in the assumption that beliefs, themselves required for moral responsibility, in turn require sentences. That assumption rests on the desire not to sever the two roles beliefs play. If beliefs not only explain behavior, but are also subject to semantic evaluation, one cannot acquiesce in a psychology which avoids sentences altogether. For how can arrays, lists, and images be true or false?

To determine whether a nonsententialist psychological theory can be adequate, we must at least recognize one when we see it. But not only does no one expect to see sentences when brain tissue is examined under a microscope; it is also the case that nonsentential representations are described sententially. When Johnson-Laird represents 'All the *A* are *B*' as:

$$\begin{array}{l} a = b \\ a = b \\ \quad\ (b) \\ \quad\ (b) \end{array}$$

he tells us that "the use of parentheses is supposed to correspond to some mental tagging of the elements to indicate that the corresponding individuals may, or may not, exist".[32] Thus, he concedes that the items '(*b*)' 'introduce' a propositional component, viz., the modal proposition that it is possible that some *B*'s are not *A*'s.

The antisententialist might retort with the familiar observation that, if we require psychological theories to be computable, they can be translated into Turing machine notation. Hence, regardless of the concepts used to describe the features of the models postulated by the psychological theory, we can replace these concepts with strings of symbols. Although the resulting form of representation involves discrete symbols, they are not sentences per se.

3.2. Modularity and Emergence

Sententialists would retort that, just as a computer can instantiate a variety of languages which differ in their level of abstractness, so a human being can be regarded on one level as an instantiation of a theory which postulates operations on sentences (and perhaps a host of other representational entities). He may even be undaunted by the fact that our functional organization at some subpersonal level will not reflect the familiar intentional distinctions we make at the personal level and that, therefore, individual beliefs cannot be identified as explanatory elements of the model. At this level, there is no relatively isolable entity, a particular belief, characterized sententially, invoked to explain some utterance. The structure of thought, which accounts for a variety of phenomena, including acts of recollection and other verbal behavior, will not incorporate units which can be paired with individual intentional units like beliefs. A third-party description of a person as acting from a certain belief will not be grounded in a psychological theory which pairs such descriptions with distinct functional states, a type of theory Stich calls 'modular'.[33] The sententialist will simply argue for the invocation of sentences and beliefs at a higher level. Dennett, for example, notes the emergent character of belief ascriptions. A chess-playing program may be said by a designer to 'believe' it should get its queen out early, playing just like a person who has this belief, although "nowhere is anything roughly synonymous with 'I should get my queen out early' explicitly tokened. The level of analysis to which the designer's remark belongs describes features of the program that are, in an entirely innocent way, emergent properties of the computational processes that have 'engineering reality'".[34]

Suppose we bracket Dennett's view that attributions of beliefs and desires are instrumentally justified in terms of their predictive utility, a position which "shields ascriptions of belief and desire from most sorts of scientific falsification".[35] The chess example, interpreted realistically, may still be shielded from falsification by dint of the emergent status of the belief ascription. Stich disagrees:

> If we were to run across (what appeared to be) a person whose conversation, chess playing, and other behaviors were controlled by an enormous, preprogrammed branching list of what to do when, I think our intuition would rebel at saying that the 'person' believed that I was about to attack with my queen – or indeed that he believed anything else! Entities with innards like that don't have beliefs.[36]

Although it is true that this 'person' does not have beliefs, and that, therefore, there are constraints on the functional mechanism explaining the behavior that seems belief-inspired, the question at issue is whether *we* might turn out not to have beliefs. Unless we are totally ignorant of our underlying mechanisms, this is different from the question of the possibility of simulating our behavior by a mechanism not informed by beliefs. The point is that we can be reasonably confident that we are significantly dissimilar from Stich's 'person with apparent beliefs'. We can, in other words, distinguish ourselves from the clear cases of behavior-apparently-inspired-by-belief which turn out to be illusory. We know enough about the inner workings of our mind to conclude that a far more sophisticated demonstration, one clearly unavailable now, is required to establish that *we* are devoid of intentional states.

Stich's example does, however, point to the conclusion that, even if we treat belief as an emergent feature, its postulation on the basis of appropriate behavior is not compatible with just any internal account. Hence, no *a priori* considerations preclude the emergence of a successful nonmodular theory which will clash with the assumption that there are beliefs.

The term 'clash' is used because the success of such a theory may not affect the continued belief in belief since the assumption that belief is emergent defuses the failure to map distinctions in intentional discourse onto elements of the model provided by the theory. The key question, unfortunately a nebulous one, concerns the specification of the stage of inquiry after which it is mistaken, obfuscatory, or in some way or another unacceptable to continue to posit beliefs as explanatory scientific features. A person attracted to Dennett's appeal to emergence may at least demand more stringent conditions on the elimination of belief than the absence of a mapping from sentential assents to fundamental units of the mental mechanism. Although the demand may be reasonable, the tricky issue is the delineation of the point beyond which it becomes unreasonable. What are the features of a nonmodular theory which has jettisoned the concept of belief?[37]

3.3 Sentences and Information

The vagueness of this issue is exacerbated by vagueness pertaining to the very concept at the heart of the dispute, that of a sentence. Consider Churchland's speculation that "any declarative sentence to which a speaker would give confident assent is merely a one-dimensional *projection* (. . .) of a four- or five-dimensional 'solid' that is an element in his true kinematical state (. . .). Being projections of that inner reality, such sentences do carry significant information regarding it and are thus fit to function as elements in a communication system".[38] But they reflect only "a narrow part of the reality projected".[39]

> (. . .) a form of language far more sophisticated than 'natural' language, though decidedly 'alien' in its syntactic and semantic structures, could also be learned and used by our innate systems. Such a novel system of communication (. . .) would enhance epistemic evaluation (. . .) since it would reflect the underlying structure of our cognitive activities in greater detail than does natural language.[40]

The latter portion of this quotation indicates, as do other statements of Churchland, that the envisaged displacement of folk psychology will incorporate a new orientation towards human cognition, one which retains notions like communication, information, and epistemic evaluation. Whether or not 'belief' is associated with too many eventually outmoded associations and theoretical excrescences to survive this journey, 'cognition' and 'epistemic evaluation' reappear unscathed. Since these conditions continue to play the causal roles expected of them, since, that is, we act on the basis of our epistemic evaluations, my intuitions tell me that we will also retain the idea that human beings are responsible agents. If this is a mistake, the reason will have to be drawn from more specific features of these theory sketches than have been provided. As it stands, it appears we will retain the idea of a bit of information and the proposition-like attitudes associated therewith.

Churchland is here advancing an idea, now commonplace, to the effect that there is an underlying representational structure, one which some believe is present in nonlinguistic organisms, that 'projects' sentences in the case of organisms which *do* have languages. And since, as we noted in section 1, we can never entirely divorce belief from assent, elements of that representational structure *might* be accorded the status of beliefs so long as they serve the appropriate causal role in relation to assent. But then, in the absence of a general theory of syntax, one designed to capture the sentence-like elements of this representational structure or code, we have no clear conception of the dispute. So, if we permit Dennett's emergence *and* if we construe Churchland's remarks as not unfavorable to sententialism (on an expanded account of 'sentence'), the only enemy of the 'sententialist' would be a rabid anti-representationalist.

These thoughts cohere with some conclusions drawn in section 2. We noted there that agents who express their beliefs linguistically must be supposed to have more beliefs than neurological states. If neurological states are associated with *sets* of beliefs, then, it is plausible to suppose that a neurophysiologically informed psychology will in all likelihood be nonmodular. That is, if the connections between neurological and psychological units are oblique, a functionalist account will be forced to postulate individual beliefs at a high level of abstraction or be forced to forego them altogether. (The Churchlands are at least right to remind us that there is a limit to the viability of a functionalist account which ignores the neurophysiological parallels between intelligent behavior in humans and intelligent behavior in other species.)

3.4. Eliminativism

If the vagueness of the issues has permitted the 'neosententialist' to evade refutation, we should remind ourselves that some are dubious of representationalism in general. Winograd, for example, endorses Maturana's point that representational accounts may be projections onto systems which can ultimately be understood mechanically and Churchland's point that we have no right to be anthropomorphic in psychology.[41] I do not want here to speculate on the possibility that eliminative materialism will eventually be victorious. One cannot provide *a priori* guarantees that psychology will always underwrite moral agency.

A less extreme version of eliminativism would permit representations, but not beliefs. Stich cites some tentative results which suggest the falsity of certain empirical presuppositions of belief-guided behavior. Data from the literature on dissonance and self-attribution support the speculation that the cognitive system which mediates nonverbal behavior is isolable from the system whose output is verbal expression and explanation. The processes which underlie behavioral responses are inaccessible to the subject whose description and account of the causally relevant processes and states are error laden, being grounded in (crude, inaccurate) socially shared theories about psychological explanation rather than insight into the actual intermediary between stimulus and response.

Stich's response to the above is as follows: "it is a fundamental tenet of folk psychology that *the very same* state which underlies the sincere assertion of '*p*' may also lead to a variety of nonverbal behaviors".[42] But if these findings can be generalized,

> (. . .) then states similar to the one underlying our own ordinary utterance of '*p*' do *not* also participate in the production of our nonverbal behavior (. . .). If we really do have separate verbal and nonverbal cognitive storage systems, then the functional economy of the mind postulated by folk theory is quite radically mistaken. And under those circumstances I am strongly inclined to think that the right thing to say is that *there are no such things as beliefs*.[43]

Although the evidence currently available is meagre in relation to the dual systems hypothesis, the case against belief is significantly flimsier than we are led to believe. As far as I can see, the results, which by the way corroborate many earlier experiments, indicate that human beings can be very wrong in explaining their own behavior, and produce corrigible and, therefore, occasionally erroneous reports of their own mental states and processes. Each of these results has been advanced numerous times by philosophers and psychologists from various traditions and with different overall programs.[44] The collective weight of these conclusions is a far cry from the demise of belief. I have two reasons for saying this.

The weaker reason is that the explanation of the behavior of the subjects, including the very dissonance that grounds the radical inferences, is studded with the notion of belief. "If subjects are led to behave in ways they *find* uncomfortable or unappealing and if they do not have what they *take* to be an adequate reason for enduring the effects of this behavior, then they will come to *view* the behavior or its effects as more attractive (italics mine)."[45] Thus, we are owed some recasting of all of this. Now I have no objection to bootstrapping and am openminded, therefore, about the prospects of rewriting. Stich *suggests* a cheap way of doing it when he describes the hypothesis under consideration as the view "that verbal and nonverbal behavior are subserved by different belief-like cognitive subsystems".[46]

But my principal objection is that our *present* concept of belief has long been recognized to be linked only tenuously to behavior, both verbal and nonverbal. Numerous philosophers of mind during the last twenty years or so have pointed to the difficulty of characterizing belief in terms of a dispositional statement whose consequent describes behavior. Everyone knows that the antecedent would have to be indefinitely long or hedged in one way or another, that it might be advisable to shift to probabilistic or tendency judgments, that some antecedent clauses will refer to states, e.g., desire, as complex as belief. My point is that any discrepancy between verbal and nonverbal behavior can just as easily be explained in terms of these considerations, which do not force a rejection or bifurcation of the concept. If I want an apple, I will not do *anything* to get one. Believing that one is in the next room, I may or may not go to get it, depending on a variety of factors. Thus, some behavioral manifestations depend on very different conditions from others and the verbal expression of a belief may consequently depend on a distinctive set of considerations, e.g., perhaps awareness, not required for nonverbal manifestations. Just because the mechanisms underlying verbal production incorporate distinctive paths through the mental machinery does not in itself constitute a case that a univocal concept of belief is not at some node common to the pathway leading to nonverbal response.

The failure of dispositional analyses of belief *as* analyses of the 'ordinary' concept can explain our hesitancy to accept incorrigibility (if I believe that I believe that *p*, then I believe that *p*) or transparency (if I believe that *p*, then I know that I believe that *p*), the traits *re*rejected by dissonance theorists. Believing is one thing, recognizing or affirming the belief another, and there can be interferences which undermine the usual associations between the two. The acknowledgement of such matters has never to my knowledge been advanced as a stipulation concerning a new concept of belief, but rather as features, perhaps disputable ones, of our old concept.

3.5. The Recent Situation in Cognitive Psychology

Philosophers, with their traditionally polemical temper, may fail to have noted the eclectic character of many cognitive psychologists who do not have the same axes to grind. For example, central to Johnson-Laird's theory of mental representations is the notion of mental models, explicitly nonpropositional constructs. But he insists on a level of processing which is propositional. We represent utterances (sentences) propositionally and these propositional representations are the partial basis for the construction of mental models. (Mental models are nonpropositional because, like images, they must be far more specific than a system of propositions need be. Also, as structural analogues of the world, they do not have an arbitrary structure, as propositions do.)[47]

Those, like Kosslyn, who argue that images are a distinct sort of mental representation, do not preclude the existence of propositional elements, even in the cognitive functioning of imagery.[48]

Although Anderson invokes trees, schemas, images, and linear orderings, he is not an enemy of propositions. "Propositions represent the atomic units of meaning and can be used to represent the meaning of sentences and pictures. The interconnections among propositions define a network that can be profitably used to understand memory phenomena."[49] What we remember, meanings rather than syntactic objects or, in the case of pictures, physical details, are stored *as* propositions.

Minsky's views cannot be interpreted unless one is permitted to include propositions as elements of a model of the mind. What he objects to in the 'logistic approach' is not the use of propositions per se, but rather other elements. "Perhaps the better part of knowledge is not propositional in character, but interpropositional. For each 'fact' one needs meta-facts about how it is to be used (. . .). But though the system can make deductions implied by its axioms, it cannot be told when it should or should not make such deductions."[50]

Of course, even if the current scene is not as rabidly anti-propositional or anti-sentential as some would have us believe, the cognitive psychology of the future, inspired by developments in neuro-physiology, is likely to look very different. Short of the radical transformation envisaged by the eliminativist, however, I have not been able to uncover a strong case against belief.

NOTES

[1] B. Berofsky (1987) *Freedom from Necessity: The Metaphysical Basis of Responsibility*, Routledge & Kegan Paul, chap. 2.

[2] Davidson has criticized various attempts to link intention with belief, but none affects this claim ('Intending', *Essays on Actions and Events*, Clarendon Press (1980), pp. 91–96). He notes, for example, that an intention to do *A* does not imply a belief one will do *A*, and that one can do *A* intentionally without knowing (or even believing with any confidence) that one is doing *A*. My claim, distinct from these, is that *B* cannot be a person's intention in doing *A* unless he believes *A* has some chance of eventuating in *B*. If *B* is important and *A* is the only possible means to *B*, a person may do *A* with *B* as his intention, even if, due to the unlikelihood of success, he does not believe he will do *B*.

[3] Some philosophers actually endorse this approach for cognitive psychology. Stalnaker, for example, goes even further, recommending that beliefs be identified with propositions, abstract entities related to, but identical to neither sentences nor neurological states. (Robert Stalnaker (1976) 'Propositions', *Issues in the Philosophy of Language*, A. MacKay and D. Merrill, (eds.), Yale University Press, pp. 79–81.)

[4] Daniel Dennett (1978) 'Brain Writing and Mind Reading', *Brainstorms*, Bradford, p. 47.

[5] John Vickers (1969) 'Judgment and Belief', *The Logical Way of Doing Things*, Karel Lambert, (ed.), Yale University Press, pp. 39–64; Ronald de Sousa, (September 1971) 'How to Give a Piece of Your Mind; or the Logic of Belief and Assent', *Review of Metaphysics* **XXV**, pp. 52–79; Daniel Dennett, 'How to Change Your Mind', *op. cit.* (note 4), pp. 300–309.

[6] .See **op. cit.** (note 4), p. 48.

[7] See Dennett, **op. cit.** (note 5), pp. 306–307.

[8] Christopher Cherniak (April 1981) 'Minimal Rationality', *Mind* **XC**, No. 358, pp. 163–164.

[9] Colin McGinn (1982) 'The Structure of Content', *Thought and Object*, Andrew Oldfield, (ed.), Clarendon Press.

[10] *Ibid.* p. 211.

[11] Hilary Putnam (1975) 'The Meaning of "Meaning"', *Language, Mind and Knowledge: Minnesota Studies in the Philosophy of Science*, vol. **VII**, Keith Gunderson (ed.), University of Minnesota.

[12] Tyler Burge (1979) 'Individualism and the Mental', *Midwest Studies in Philosophy*, vol. **IV**, (eds.) Peter French, Theodore Uehling, Jr., Howard Wettstein, (eds.) University of Minnesota.

[13] Stephen Stich (1983) *From Folk Psychology to Cognitive Science: The Case Against Belief*, MIT, pp. 63–64.

[14] Jerry Fodor, 'Methodological Solipsism Considered as a Research Strategy in Cognitive Psychology', *RePresentations*, MIT.

[15] 'The Structure of Content', p. 215.

[16] Hartry Field (July, 1977) 'Logic, Meaning, and Conceptual Role', *The Journal* of Philosophy **LXXIV**, No. 7, pp. 379–409.

[17] Daniel Dennett, 'Beyond Belief', in Oldfield, *op. cit.* (note 9), pp. 38–52.

[18] Tyler Burge, 'Other Bodies', in Oldfield, *op. cit.* (note) 9, pp. 116–118.

[19] This is true on the assumption that descriptions of behavior are not intentional so that they are not infected by the same problems. One must describe Robinson's behavior not as the seeking of relief from arthritis. For a discussion of the nature and importance of this requirement, see Stich, *op. cit.* (note 13), pp. 165–170, 194–198.

[20] Donald Davidson, 'Mental Events', *op. cit.* (note 2), pp. 207–227.

[21] René Descartes (1979) *Meditations on First Philosophy*, (tr. Donald A. Cress), Hackett, pp. 20–22.

[22] This 'property' might represent a combination of many other properties.

[23] See *op. cit.* (note 13), p. 151.

[24] Hartry Field (July, 1978), 'Mental Representations', *Erkenntnis* **XIII**, No 1, p. 44.

[25] Hilary Putnam, (1983) 'Computational Psychology and Interpretation Theory', *Realism and Reason: Philosophical Papers*, vol. **III**, Harvard University Press, pp. 139–154.

[26] For a related critique, see Alvin I. Goldman (1986) *Epistemology and Cognition*, Harvard University Press, pp. 173–177.

[27] Donald Davidson (1986) 'Judging Interpersonal Interests', *Foundations of Social Choice Theory*, J. Elster and A. Hylland, (eds.) Cambridge University Press, pp. 195–211.

[28] Perhaps it follows that I must also suppose most of your beliefs are true. But talk of truth complicates matters unnecessarily. It certainly does not follow from these considerations that most of your beliefs are true. There might be – or I might believe there is – a Cartesian demon who deceives you and me about most matters. To be sure, Putnam has raised issues that bear upon such skepticism. See his (1981) *Reason, Truth, and History*, Cambridge University Press, pp. 1–21. For a similar criticism of Davidson, see Goldman, *op. cit.* (note 26), pp. 175–176.

[29] See *op. cit.* (note 27), p. 206.

[30] Philosophers have finally come to realize that we are not nearly as rational as they have portrayed us. See, for example, *op. cit.* (note 8).

[31] Here again, Goldman agrees. See *op. cit.* (note 26), pp. 173–177.

[32] P.N. Johnson-Laird (1983) *Mental Models*, Harvard University Press, pp. 163–164.

[33] See *op. cit.* (note 13), pp. 237–238.
[34] 'A Cure for the Common Code?' *op. cit.* (note 4), p. 107.
[35] See *op. cit.* (note 13), p. 243.
[36] *Ibid.* pp. 244–245.
[37] For a defense of the view that we are unlikely to construe the nonmodular theories which emerge as necessitating the abandonment of belief, see Terence Horgan and James Woodward (April, 1985) 'Folk Psychology is Here to Stay', *Philosophical Review* **XCIV**, No. 2, pp. 197–226.
[38] Paul Churchland (February, 1981) 'Eliminative Materialism and Propositional Attitudes', *The Journal of Philosophy* **LXXVIII**, No. 2, p. 85.
[39] *Ibid.*
[40] *Ibid.* 86.
[41] Terry Winograd (1981) 'What Does It Mean to Understand Language?' *Perspectives on Cognitive Science*, Donald A. Norman, (ed.), Ablex; the observation by Patricia Churchland appears in 'A Perspective on Mind-Brain Research' (1980), *The Journal of Philosophy* **LXXVII**, p. 207.
[42] See *op. cit.* (note 13), p. 231.
[43] *Ibid.*
[44] In defense of the corrigibility of causal judgments in psychology, I defended these claims, citing psychological literature outside the Freudian framework, in (1971) *Determinism* Princeton University Press, pp. 113–114; see also Daniel Dennett, 'Toward a Cognitive Theory of Consciousness', *op. cit.* (note 4), pp. 164–173.
[45] See *op. cit.* (note 13), p. 232.
[46] *Ibid.* p. 233.
[47] See *op. cit.* (note 32).
[48] S.M. Kosslyn (1980) *Images and Mind*, Harvard University Press.
[49] J.R. Anderson (1985) *Cognitive Psychology and Its Implications*, Freeman, p. 133.
[50] Marvin Minsky (1981) 'A Framework for Representing Knowledge', *Mind Design* John Haugeland (ed.), Bradford, p. 124. Minsky's concern that propositions are not the units of meaning does not ipso facto make him an enemy of propositions. Philosophers, e.g., Quine, have challenged the view that sentences are the unit of meaning without being taken as denying that there are sentences or that we think in sentences.

Department of Philosophy
Columbia University, New York, U.S.A.

CHRIS MORTENSEN

MENTAL IMAGES: SHOULD COGNITIVE SCIENCE LEARN FROM NEUROPHYSIOLOGY?

1. INTRODUCTION

The debate over the status of mental images has been of perennial interest to philosophers; and it has recently taken a decidedly experimental turn, but with battle lines being drawn similar to those in the earlier phase of the debate. This paper aims to argue that the recent experimental evidence has been somewhat one-sided, deriving as it has pretty much from cognitive psychology. It will be argued that from that viewpoint a fairly natural theoretical elimination of mental images is possible; but it will also be argued that to do so is to close one's eyes to theoretical considerations of a different kind, deriving from the discipline of neurophysiology. It will be speculated that, when these neurophysiological considerations are brought in, it might be possible to resurrect at least some parts of the mental image story, while remaining within a broadly physicalist framework.

One quick disclaimer. The argument of this essay leans heavily on propositional attitudes of various kinds, particularly beliefs. It is, however, not intended to take sides concerning the theoretical dispensibility of beliefs, as instanced by Dennett's views for example. Only this much is assumed: some minimal theoretical and explanatory use in attributing beliefs and other propositional attitudes to a system. This is, I take it, compatible with various degrees of realism about propositional attitudes, and various views on how far down the pyramid of software breakdown items worth calling beliefs might continue to appear. It is this sense in which I hope it is possible to avoid taking sides on the matter.

2. NONVERBAL PROPOSITIONAL ATTITUDES

I need to distinguish between different kinds of content which beliefs and other propositional attitudes can have. The principal distinction I urge is between beliefs whose content is wholly expressible in words (*She believes that Canberra is the capital city of Australia*), and beliefs whose content is not wholly so expressible. At this point the sceptic

Peter Slezak and W.R. Albury (eds.), Computers, Brains and Minds, 123–136

invites us to tell him what one such belief might be, and we must not fall into the trap of trying to describe its content wholly in words. But we can show such beliefs without exhausting their content in words. You believe that his face was like *that*. The whole point of police identikit pictures is precisely that verbal descriptions exhaust only a tiny fraction of the information stored in us. Such contents, moreover, are generalisable both to other sense modalities (I believe that the music goes like *that* or like ) and to other propositional attitudes. (He desired Eliza Doolittle to sound like that, I intend that his face look like this.)

The philosophical advantages of admitting nonverbal beliefs are considerable. In the first place, the way is opened for dumb beasts to have propositional attitudes. This is, of course, a natural view, and one in accordance with the current push for animal rights. For instance, it might be argued that whatever has rights has interests, and whatever has interests has wants, desires and goals. But only if one holds the narrow identification of the content of propositional attitudes with verbal content should one have trouble accepting this.

Indeed, there is reason to hold that appropriate propositional attitudes are essential if there are any moral imperatives at all due to animals. Take pain: in what sense could the animal dislike the torture if it doesn't desire that *this* (pain) stop, or take avoidance behaviour because it desires that *this* not happen, for example? In fact, if animals store information at all, then the information they store is not verbal information. What they have the capacity to store is information about their environments which is important to them, and there is no reason save poor philosophy to say that such information is represented to them with a somehow verbal content, or for that matter that it could be expressed in our language. This in turn leads to a second philosophical advantage of nonverbal beliefs in humans, namely the evolutionary continuity between animals and humans, since this would suggest that we, too, store some information nonverbally. What kinds of beliefs did we have before language? The human paleocortex represents an evolutionary stage of information acquisition prior to language, which is nonetheless of great sophistication as an information-processing unit.

Another advantage of nonverbal beliefs is that their power to abolish *qualia* is almost magical. The problem is that we have information about our own states, that they seem for all the world to have qualities about which it is hard to tell some physicalist software – or hardware – tale. But if the qualities of sense data are bracketed by a belief operator, then the problem of their reification vanishes. If instead of hallucinating a geometrical figure, I simply believe that I see that figure, then there is no geometrical figure to hunt for. Unfortunately it is simply unbelievable that the sensory richness of my experience does not far exceed mere beliefs that I see a figure under a certain discription, if only because I may have no name for the figure. But bring in beliefs whose content involves *that* figure, and we seem to have as many sensory features as one would like, all neatly semantically bracketed and ontically sanitised. And so, furthermore, there is no obvious reason why such features of beliefs might not be stored as patterns of zeros and ones.

3. ACCOUNTING FOR IMAGININGS

I now want to apply the distinction of the previous section to the debate about mental images. The aim of this section will be to argue that admitting nonverbal beliefs rather neutralises the experimental evidence allegedly in favour of spatially structured mental images. In turn, this will clear the way for subsequent sections to argue that the case for mental images has been mislocated in the standard 'top-down' cognitive science approach, when one might instead look elsewhere, to neurophysiology.

I do not propose to go deeply into experimental detail, so much as to try to describe the *kind* of experiment which has been proposed[1] in favour of pictorialism, and against descriptionism. Subjects are shown or instructed to imagine something with spatial characteristics, such as a letter of the alphabet, or a map, and then instructed to perform in the imagination some spatial transformation on it, such as rotating it, or scanning it along some direction. It turns out that the time taken to perform these tasks successfully bears a simple (linear) relationship to the amount of spatial transformation required of the imagined object, for example the angle through which the figure has to be rotated. These results were taken by Shepard to suggest that imagining a shape requires that there really be a shape in existence, rather than, say, imagining a shape being a primarily verbal event, of the kind: the (verbal) belief that there is a square in existence. Otherwise, it is difficult to account for the *temporal simplicity* of the transformations of the imaginings. The theory was further developed by Kosslyn and others, who postulated a 'visual buffer' (he uses the analogy of the television screen) on which imagined items appear and are transformed in accordance with laws which take a simple form when described in terms of the spatial relationships of the buffer.

Zenon Pylyshyn strongly criticised the theory on a number of grounds. He argued that for the theory to be an explanation of the data, then the laws governing the transformations should derive from the physics of the medium, which is difficult to credit since the representation of information is entirely digital. He also argued that the idea of a visual buffer is at odds with the evident cognitive penetrability of the imagination: background beliefs affect what is imagined and the laws governing transformations. He claimed that one could account for the experimental data by supposing that, for example, instead of imagining a rotating letter of the alphabet being a matter of a letter of the alphabet actually rotating (in some visual buffer), it rather involved *imagining what it would be like to see the rotating letter*. This conveniently cauterises the ontic commitment as before; and in addition one can draw on one's experience of what seeing such events is like in order to explain the time laws obeyed by the transformations.

I do not propose to discuss Pylyshyn's arguments in detail, though I will indicate where later points bear on them. But a concession of Kosslyn's to Pylyshyn opens the way for a general argument for Pylyshyn's conclusion. In one place,[2] Kosslyn makes out that his visual buffer is not to be understood, so to speak, as a *real* spatial array, so much as a convenient (and theoretically fruitful) *software* story. Now this would seem to be fatal for any kind of realism about images with spatial properties, since it allows that in the end the elaborately conceived visual buffer is just so many ones and zeros again. And it would also seem to be a concession not at all in keeping with the spirit of pictorialism, as assimilating images to real pictures.

We can come at this point from another direction, that of nonverbal propositional attitudes. Suppose that underlying attitudes such as believing that one sees a rotating letter, or imagining a rotating letter, is, as described before, a nonverbal attitude involving *that* letter. Now let us ask what identifies the content of a belief, verbal or otherwise, as being the content that it is? Some people have focussed on the deductive relations between beliefs, while others have objected to the priority this gives the proof theory over truth theory in natural semantics. At any rate, it isn't apparent what *deductive* relationships are constitutive of the belief that his face is like *that*. This is not to say that such beliefs bear no systematic relationships to other nonverbal beliefs. It's hard to imagine what such a belief could be if it were not surrounded by potential judgements about *similarity* between the face and others, for example. More general would seem to be the fact that *quick* deductions from verbal beliefs and *easy* judgements of similarity between the nonverbal beliefs are *important* to the identity of the proposition believed, indeed more important to its constitution than the ability to deduce distant equivalents or recognise nonobvious similarities. This point might be summed up in an *Explanatory Principle of Natural Semantics*: one might expect a system to be able to perform those transformations which are describable as *simple* functions of the content which it is independently fruitful to ascribe to the system.

I am less interested to argue for this principle than to hypothesise it as useful and plausible. Let us see what it does to the results of the mental imagery experiments. I claim that if one thinks it worth ascribing the nonverbal propositional attitude of imagining a certain shape to a system (such as a human) then it should not be so surprising that the system be able to perform simple spatial transformations fairly simply on the imagined shape, *independently* of how one supposes the imagining of the shape to be realised, as something involving an actual shape or as mere software. To say that it performs certain transformations simply is to suggest, furthermore, that one should not expect *more* complex functions than we are already prepared to entertain to describe transformations; and that is to say that it should not be surprising if one finds these functions to be descriptively fairly simple at the postulated level of explanation. In short, the admission of software involving imaginings with a nonverbal, specifically spatial, content means that we should be in a position to take advantage of software involving (relatively) simple transformations of that content, so that it might not be unexpected that laws governing the ability of the system to make such transformations be simple as well.

I claim that it is possible to read both Kosslyn and Pylyshyn in ways which make them consistent with the above. Kosslyn's concession that the visual buffer is mere software should lead one to ask what kind of semantic bracketing *imagining* puts around the spatial characteristics of a visual buffer so as to remove it from ontic consideration. The natural move to make is to say that such imaginings have a nonverbal propositional content, since it was after all a verbalist reduction which Kosslyn was most concerned to combat. The transformations on Kosslyn's visual buffer are then to be seen as transformations on the nonverbal propositions imagined, and it is not unreasonable to think that simple transformations will be subject to simple laws. On the other hand, one feels that Pylyshyn's verbalism is easily dispensible in favour of a more general *propositionalism*, the virtue of which is that spatial qualia are contained within the semantic brackets of the imagination, and ontic commitments are no more than those of the hardware realisation of any piece of software. Furthermore, Pylyshyn's suggestion that the transformations in question might only amount to imagining what it would be like to see it rotating, can be seen as a special case of the Explanatory Principle of Natural Semantics. Pylyshyn is suggesting that a process which is fairly simply describable in the terms of the contents of his preferred propositional attitudes, might suffice to account for the simple laws found to govern the transformation. So it might, and the more complex the postulated process the harder it would be to believe that it follows simple laws.

To sum up this section, it has been argued that existing arguments *from cognitive science* designed to establish the existence of mental images can be thwarted by the recognition of a broader class of contents for propositional attitudes than those expressible verbally. Even so, one feels a certain dissatisfaction with this state of affairs. Is there no way that mental images *might have* been? At least one might say that real mental images are *consistent with* the findings of the experiments. Can we do no better than the tired argument that the theory without images is simpler? In the next section, I turn to what could be called speculative neurophysiology, to try to sketch a more positive role for images.

4. SPATIAL STRUCTURES OF NEUROPHYSIOLOGICAL SIGNIFICANCE

It is known from neurophysiology that visual perception proceeds by means of a point-to-point projection of the image on the retina into the primary visual areas of the cortex, specifically Brodmann's areas 17, 18 and 19. The sense in which the spatial structure of patterns of firings on the retina is preserved in the primary visual cortex is that while fairly similar spatial patterns of shapes occur in the latter, they are distorted to some extent, and in fact cut in half, in that the left half of the visual field (to the left of the retinal fovea) of both eyes appears in the left hemisphere, and correspondingly for the right. I will say more about the significance of these distortions from true later; however, for the moment we will set them aside.

I want to suggest that the shapes in the primary visual cortex might *be* a component in the nonverbal perceptual belief, or for that matter the nonverbal imagining, that there is this shape in my visual field. If one admits nonverbal propositional attitudes, one needs to ask about the hardware mechanism of their realisation. And, I suggest, it might be a simple and natural mechanism for so doing, that carbon-based lifeforms do realise *some* such subsystems, or groups of attitudes, in ways which utilise isomorphic features of the medium to contain the information. Indeed, prior to language, how *otherwise* might an organism go about representing information about shapes within itself?

Note one disclaimer. I am not saying that *all* the contents of perceptual beliefs need to be represented by the same features in the representing medium. I am suggesting it only for *some*, specifically shape and touch. It is hard to make the account work for colours which is somewhat unfortunate because of the evident relationship between shapes and colours. I will say something about sounds later.

Dennett emphasises that experiences must have a belief component which is something (causally) more than any putative image. We can agree with that here without concluding that the image is superfluous. One could say that the perceptual belief that there is a particular shape in my visual field is a complex causal structure, a causally significant component of which is that shape of neural patterns in the believer. And yet there is also a sense in which the old idea of the incorrigibility of sense data might be preserved, in that the awareness of the shape and the shape might well be *causally* indissoluble.

Situations of various sorts of imagery, for example outright hallucinations of shapes, get a natural account on this view. The situation concerning the normal perception of shapes in front of the eyes is that the shapes play a causal role sufficiently significant (and differentially significant) in the process for it to be worth calling information about *the shape* that is being stored. Turning this around, if a shape has sufficient causal importance to the occurrence of the belief that the shape is in existence, then it seems reasonable to say that one is *aware of* that shape, or knows that there exists a shape (in one's visual field) *even if* the shape is somewhere in the central nervous system rather than in front of the eyes, or on the retina in afterimaging. Consequently, one could go on to say that the philosophical elimination of shaped visual objects under conditions of hallucination, and imagination, is unnecessarily drastic, and has no better *a priori* motivation than a similar elimination of seen objects from in front of the eyes. It is not an unnatural view that human beings are skilled discriminators of shapes; so that when they claim to be aware of the existence of a shaped object when patently there is no shaped object in front of the eyes, this by no means implies that their ability to discern the existence of shapes has gone *totally* haywire. At the risk of boring the reader, let me repeat that it is not being claimed that the existence of such shapes is *logically necessary* for perception, imagining or hallucination. The claim is that this may well be how the mechanism is for us, and maybe must be for wet creatures.

A matter to make sure of, is that the shapes in the visual cortex, and indeed the visual cortex itself, are not functionally epiphenomenal in the story. One thing which neurophysiologists never tire of stressing is how little is known about what happens to the information sent on from the primary visual areas. Certainly it is sent to many places. But equally the primary importance of the visual cortex is unquestionable. Its connections with the rest of the system are too complex for it to be a mere staging point for information, as with the optic chiasm or lateral geniculate bodies. The other matter which reinforces the idea that the spatial structures in the visual cortex are not functionally epiphenomenal, is the fact that the information is not simply spatially recorded *once* in the visual centres, but is reproduced in many places. This is prime evidence that the organism values the information, and in that form. The argument here for functional significance is not the familiar AI functionalist analysis, so much as an argument from what the neuroanatomy would suggest. Thus, I claim, cognitive science has sometimes been rather too fixated on functionalist software considerations when it might well learn from the fact that we are also a piece of evolved hardware. We weren't designed by a cognitive scientist, we just grew.

The causal/functional role suggested for shapes in the visual cortex extends fairly easily to a different modality, the somatosensory system. The structure of the skin is also represented in a (relatively) isomorphic spatial structure in the somatosensory cortex, the shaped homunculi being familiar from physiology texts. The shapes in question are evidently of importance to the organism, and the mode of storage of information does not seem to be unreasonable or inefficient. So we have, too, a natural account of hallucinations concerning the shape of the skin or spatial disposition of body parts, for instance phantom limb phenomena. One might say that sensing a phantom limb is like sensing a real limb in a particular simple way, namely that in each case there is in existence a spatial item (in the CNS) which bears the *same* spatial relationship to a larger spatial structure in the CNS as the *real* limb does to the real body. And furthermore, the activated cortical structure is of sufficient causal significance to the information in question for it to be worth saying that one *knows or can know* of its existence. That much is not illusion, at any rate.

Let me also re-emphasise the speculative nature of this story. Whether it works is in the end a matter of how the hardware and the software turn out to mesh together, without doubt. My aim is rather to bring general considerations to bear to show that some such story is by no means crazy, consistent with the current state of brain science.

The idea of the contents of information being represented in nerve nets by features isomorphic with those contents does not work for some other sensed qualities, however. One might hypothesise that sounds in the head are isomorphic with sounds outside, in the sense that, say, a sound in the head is a real electrical vibration of the same or isomorphic frequency as the air vibration. Promising as it appears, this does not seem to be how sounds work. Isomorphic frequencies are to be found along the auditory nerve, but these frequencies seem to be lost at the cortex, though conceivably a more fine grained investigation would uncover them. The same hypothesis could be framed for colours, too, namely that different colour sensations are a matter of different frequencies of firing, isomorphic with different frequencies of incident light. As far as my limited information goes on this, however, colour perception is a matter of spatially distinct receptors for different colours right at the retina, rather than different frequencies associated with them; and this carries back into the cortex. It would not be surprising, of course, that colour perception follows closely whatever is responsible for shape perception since colours are perceived to be spatially extended. But it does mean that representing input by causally exploiting natural isomorphisms with that input to store information might work for some kinds of information and not for others.
tion and not for others.

It should be pointed out, too, that from the very first Jack Smart understood the essence of this.[3] He claimed that what we know about the processes inside our skins when we utilise introspective information about them, was a series of topic neutral properties. He emphasised that a judgement of similarities between interior items was entirely compatible with those items being wholly physical in character, and included judgements about the waxing and waning of inner processes as well. It can be said on the present account that a judgement of similarity between inner states is closely connected with there being relatively simple transformations between those states as judged from the point of view of whatever propositional contents it seems explanatorily worthwhile to attribute to the system. Of course there was always the problem for Smart of having to say that one made judgements of similarity *in different respects* and how to account for this extra element. But here one might speculate that the respect in which the (false, phantom) nonverbal belief that I have a leg resembles the veridical belief that I have a leg, is that in both cases I know there to be in existence an item bearing the same spatial relationship to an inner neural structure, as legs do to bodies. This makes way for specific respects. Waxings and wanings get a plausible treatment too. One might allow that there is more than one way that a process could wax, either by increases of amplitude of currents, or by increases of frequency. Each might be the causally significant feature in different processes, to the point of being *constitutive* of the process of nonverbally apprehending that something (of other character) is waxing and waning within me.

Something has to be said about the previously mentioned *divergences* of the shapes in the visual system from the shapes present at the retina, specifically that retinal shapes are distorted, and in some instances cut in half. Here I think it is reasonable to inject a healthy note of eliminativism. The aim of the exercise has been partly to restore a note of veridicality to our perceptions of ourselves, but this is a battle fought in the context of widespread and enthusiastic embrace of the strategy of denying the reliability of introspective information, and denying the existence of items such as images unless they are in accord with respectable theory. Sophisticated as the central nervous system is, it is after all a fairly crude discerner of the features of the world both inside and outside our skins. It gets some things systematically wrong, and it fails to discriminate others, or fails to discriminate differences between them. One might therefore say that the brain simply *fails to perceive* that the cortical shape picturing the exterior shape is in halves. At least in a properly functioning perceiver, the causally relevant feature is the shape as *joined together* – another way of saying that the brain 'joins the shapes together'. Similarly, the brain 'corrects for the distortions' when it produces beliefs about shapes, which is to say that there is a systematic (simple?) transformation from the causally significant actual shapes, to those present to consciousness i.e. which one knows to be in existence. In normal perception this is knowledge of the external world, while in mistaken perception something remains known true, namely whatever features are *preserved* in the point-to-point retina to cortex map.

Let me conclude this section by observing that this view of the way information is stored in the central nervous system of various wet evolved animals, in turn makes the starting point for nonverbal propositional attitudes easier to come at. Contrary to Pylyshyn, there is simply no call to assume a common 'language' at any reasonably informative level of software between various information-relevant subsystems of the brain. I take it, by the way, that this is consistent with a Fodorian Language of Thought, *if* that thesis is construed as the claim that those propositional attitudes whose content *is* verbally expressible, are represented fairly specifically in structures with fine differences corresponding to syntactic and semantical features of the verbal content. The Language of Thought might even get some indirect support here, in that the brain is prepared to represent certain aspects of propositions *fairly specifically* in quite precisely differentiated isomorphic causal structures. My point is, though, that it might be quite sensible for Nature to represent different informational capacities quite differently, with the causal linkages between subsystems with different languages being *at best* translations of various degrees of crudity. Furthermore, one can discern various degrees of relative autonomy in different neural structures, which is at least suggestive of the idea of distinct *types* of propositional content. The most obvious example is the distinguishability of the *kinds* of sensory modality, but we might also note the autonomy of the sympathetic and parasympathetic nervous systems from the central nervous system.

5. THE AI FALLACY

In this section, I want to state and reply to an objection which I believe seduces many philosophers into thinking that an AI-oriented approach is the right one for cognitive science. The objection is that the causal effectiveness of neural networks is in the end carried only by the *connectivity* of those nerve nets. Preserve only the circuit connections and you preserve the system's actions, potentialities, and discriminatory ability toward its environment. Replace the nerve connections with wires and the neurons by bits of silicon, while keeping the inputs and outputs the same, and you have a system in an identical state of mind. But it is possible to preserve connectivity while destroying any special spatial features of perceptual subsystems. So it is possible to mimic the software right down to machine code by a system with entirely different *spatial* neuroanatomy. Consequently the spatial characteristics of the structures in which spatial information is stored cannot have the functional importance claimed in the last section, which would be necessary for them to be components of both the causal and the propositional aspects of the system.

It's a seductive argument, but one which, I think, precisely ignores the question of what constraints on a lifeform might derive from the fact that it is both evolved and carbon-based. Consider first the retina. The argument I have just stated would seem to prove too much about the retina, namely that the spatial structure of the image at the retina is functionally irrelevant to, say, afterimaging, or for the matter to normal perception. It is not that a receptor of the spatial features of the environmental *has* to capture them with spatially similar features, since radar does not operate like eyes. Pull the cells of the retina apart while preserving their connectivity, and presumably you preserve afterimaging. But it is not apparent how there might be an animal like this in its *natural* state. It is not apparent that such cellular architecture would be possible in a DNA-based system. Of course, there is the further point here that pulling the *receptor* cells of the retina apart would result in the wrong shapes, or no shapes, being perceived from the external world, since a shaped electromagnetic input would fail to produce the response necessary to store information about its shape. But the general point remains that it might be an extremely efficient way, perhaps the only cytologically possible way, for information to be transferred to and stored in structures deeper within the brain, that it be done in such a way that the spatial features of the nerve network be significantly one way rather than another. One might speculate that a fairly constant *axon* length for specific kinds of nerve cells, say those of the optic nerve, together with a requirement that there be a systematic number of nerve cells in the optic pathway (actually, two), more or less *forces* it to be the case that information is projected from the retina into the visual cortex in a point-to-point projection. In no other way could the information be received and stored.

Nor would it be so surprising if it were an extremely efficient way for a pre-verbal animal to store information about shapes, both exterior shapes and skin shapes, in shaped information. The efficiency here might well be energy efficiency in the transfer of information, as well as whatever cytological constraints there might be. Needless to say, it might not be the *most* efficient way of doing things, even with cells, but that is only to say that evolution takes funny twists and turns.

There is a further reason for being suspicious of the pull-apart argument. This is that the brain evidently makes heavy use of interference patterns between broad groupings of cells, and plainly it is usefully redundant to do so. But interference patterns will be seriously affected by any transformation which leaves connectivity unaffected but changes spatial distribution of cells and thus axon and dendrite length. Take the simplest of examples, the bundle of two-axon fibres which is the optic nerve responding to a changing visual scene. Distort the relative lengths of the fibres sufficiently and one inevitably changes the time distribution of the patterns forming in the visual cortex. With sufficient distortion there need be no time-slice which contains a pattern even remotely like the shape presented at the retina. This argument even applies to silicon hardware: since it takes time for a pulse to go down a wire, a change in the geometry of the connecting wires (the shape of the chip) changes the ability of the system to operate in concert. We must conclude again, then, that the kind of hardware might place considerable constraints on the nature of information and the characteristics of its representing system.

The idea that since cellular connections are the bottom line of the causal story, all that matters to cognitive science is top-down functionalist AI software, goes with the idea that if a certain functional account can be given of a particular cognitive system, that *must* be how humans do it. I want to dub the latter the *AI fallacy*. What is wrong with the AI fallacy is what has been argued in this paper, namely that the problem of physically realising an information-storing system is seldom tackled by AI researchers, and that realising such a system in evolved cellware might place quite definite constraints on the kinds of information stored and the characteristics of the storage system. An older mistake is a special case of the fallacy. Considerable philosophical effort was expended in trying to show that Direct Realism was a logically possible theory of perception. The fallacy is committed if one argues from this to the conclusion that hence the Representative theory of perception was false of humans and that hence there are no sense data. But from the fact that some system does it one way, it does not follow that a different system might not do it another way. We fail to take advantage of all the information available to theory if we ignore the neurophysiological issue of what kinds of information could possibly be stored in us, and how it might be stored.

6. CONCLUSION

Little remains to be said but to emphasise again the speculative nature of the present argument. It is to be viewed in the way one might present a logically possible counter-example to an argument, in this case the argument which dispenses with distinctively spatial mental images. My counter-claim has rather more force than the merely logically possible, since it seems to me that there are several pointers from neurophysiology which give the speculation some plausibility. At the very least, they show that cognitive science would be unwise to ignore the findings of neurophysiology.

NOTES

[1] Especially by Shepard, Kosslyn and others, and criticised by Pylyshyn. See N. Block (ed.), (1981) *Imagery*, MIT.

[2] *Ibid*. S. Kosslyn (1981) 'Medium and Message in Mental Imagery', p. 215.

[3] J.J.C. Smart (1959) 'Sensations and Brain Processes', *The Philosophical Review* **LXVIII**, pp. 141–156, and (1963) *Philosophy and Scientific Realism*, Routledge.

Department of Philosophy
The University of Adelaide, Australia

PETER SLEZAK

HOW *NOT* TO NATURALIZE THE THEORY OF ACTION

1. INTRODUCTION

This paper is really written for my small son. When he was about six years old he received one of those anatomy books in which different parts of the body pop up at you in the most remarkable ways. Here he first learned of the way in which our limbs are moved by muscle contractions, and how the muscles, in their turn, are stimulated by the nerves and so on. He had evidently been thinking about this, and one day as we were walking in the company of a non-philosophical friend, he asked – a propos of nothing – "Daddy, if the brain makes the muscles work, what makes the brain work?"

Philosophers at least, will appreciate why I had mixed feelings about being asked this question in public, as it were. On the one hand, I felt a certain parental pride at this display of philosophical precociousness: Of course, this question is just the classical problem of the Will.

On the other hand, however, the pride I felt was a bit hollow. I also felt a certain embarrassment at having to admit that some grown-up people spend a lot of their time worrying about this very question.

Most recent among these is Myles Brand, whose recent book *Intending and Acting* is an extended attempt to find the proximate causes of action and thereby an answer to my son's question. Thus, in focussing my discussion on Brand's work, I offer an analysis which is almost entirely destructive criticism; however, I see this as having an important and broader, constructive purpose, namely, as a *propaedeutic* and *prolegomenon* to a properly naturalized theory of action. In other words, we might get it right if we understand how Brand gets it so wrong. In particular, with the current lively interest in the fate of folk psychology in a mature cognitive science, the problem of action serves as an instructive case study. Brand articulates a strong version of the view which has been implicit in most philosophy of action, giving a pre-eminent place to the precepts of folk psychology.

A distinct issue which is also illuminated indirectly in the course of examining Brand's book is the fate and role of philosophy itself. The much vaunted continuity between science and philosophy can also be construed as the elimination of philosophy through depriving it of any

Peter Slezak and W.R. Albury (eds.), Computers, Brains and Minds, 137–166

distinctive role or *sui generis* contribution. A trend toward naturalization in the *practice*, as distinct from the pronouncements, of philosophy entails the demise of the philosopher as dilettante and supernumerary commentator. This is, after all, only to take seriously Quine's dictum that epistemology is subsumed as a proper part of psychology. If the discipline of AI can be seen in terms of the philosopher's armchair drawn up to the computer console – in Dennett's nice image, then the imperative to don a lab coat is hardly less relevant and compelling. Of course, the Churchlands have already taken just this logical step out of the armchair into the neurophysiology lab, and already there is little in their recent work which is recognizable as philosophy of the kind I had been taught.[1]

Though it has been completely ignored by Brand, as indeed by the rest of contemporary philosophy in spite of professing a naturalized approach, I will indicate that it is in the domain of action in particular that there is much to be learned from the neurosciences.

It is perhaps worth observing in passing, that the neurophilic approach to philosophy of the Churchlands, while foreign to the recent tradition of philosophy, would nevertheless have been perfectly familiar to Descartes, who has been much misrepresented in this regard, especially in the analytic tradition. The innovation suggested by the idea of 'neurophilosophy' in Patricia Churchland's[2] sense is actually Descartes' own approach to be seen exemplified in his *Treatise of Man* and *Dioptrics*. Thus, on one recent account[3] Descartes is described less as a 'philosopher' in the recent sense, than as a practicing scientist who happened to write a couple of relatively less significant 'philosophical' works in addition to his physiological and other scientific treatises. (See the 'Introduction' to this volume.)

2. NATURALIZING ACTION THEORY

Even before the rise of the current naturalistic ethos in philosophy, it was always far from clear that philosophical action theory could constitute an autonomous enterprise in the way that epistemology was construed. The foundationalist *motif* which gave epistemology its alleged priority and independence is absent from action theory, and without it, the *sui generis* character of the problems of action was perhaps never defensible nor, as far as I am aware, was it ever explicitly defended in practice.

Accordingly, we saw the awkward contortions (such as those of Chisholm or Danto which were typical) in which a posture was sought at once consistent with the deliverances of a deterministic, nomological science, and yet at the same time, preserving some transcendent metaphysical niche for the puzzles of human action. I have elsewhere argued recently[4] that these puzzles could be seen to be spurious on certain plausible empirical assumptions, and correspondingly, the philosophical domain or problem space is assimilated to that of natural science.[5] The characteristic implicit assumption of an extra-scientific problem space entailed a logical geography which was the conceptual analogue of the impossible figures of Maurits Escher: I will suggest that, like his drawings, many of the concepts and theories of the philosophical literature cannot be mapped onto any plausible account of empirical reality. But details aside for the moment, if the answer to my son's question is not be found squarely within empirical science, then where?

All this should appear more or less obvious now in the wake of the naturalistic spirit which pervades the cognitive revolution. But, it is perhaps typical of revolutions, that there remains a band of reactionaries, philosophical Somozistas, in the jungles of action theory: these are the volitionists, essayists and the like; the 'conationist Contras'.

Continuing this metaphor and adapting the familiar slogan slightly, we may say that the connection between interpreting the world and changing it is precisely the connection between cognition and conation. It's no news that the theory of knowledge or interpreting has been transformed by the cognitive revolution, and it is remarkable, then, that the closely allied theory of action has remained untouched by these developments. This cognitive revolution which has come to dominate philosophy of mind and epistemology arguably began with the work of Miller, Galanter and Pribram.[6] Twenty five years later, the ideas set forth in this classic work are still being actively discussed and have become the foundations of much work in psychology. Seemingly untouched by these developments, however, is the philosophy of action which has remained firmly ensconsed in the armchair and aloof from the intellectual ferment in AI, psychology and neuroscience labs. An acute irony derives from the fact that the focus of concern for Miller et al. was the understanding of human action.

Brand's book appears at first sight as a welcome, if belated, attempt to attend to scientific developments of the past and the present, and to assimilate them to the philosopher's special concerns with human action. Whether such an assimilation is possible, as Brand conceives it anyway, is a question to be raised in this study. In particular Brand devotes an important section of the book to a consideration of the Miller, Galanter and Pribram idea of TOTE schemata and to the earlier seminal ideas of William James in addition to some more recent work on the computer simulation of planned, intentional action. He thereby appears to redress the imbalance which has prevailed through the neglect of science in philosophical action theory.

Thus, the aspirations of the book are high, and the goals which are boldly proclaimed in the preface are nothing less than that of naturalizing the philosophy of action. Brand observes that in the 1970s "stagnation set in" and "positions hardened, epicycles were added, and progress was slowed. Interest waned".[7] By contrast, Brand asserts:

> The goal of this book is to usher in the next – and third – stage of philosophical action theory. This stage is similar to the second in that the focus is systematic theorizing about human action. But it differs from its predecessors in being continuous with *non*philosophical work on human action. Through the second stage philosophers generally ignored the scientific study of action. Of course, passing mention of related psychological theories was made; but there was no serious attempt to integrate philosophical action theory with the relevant scientific theories. My contention is that future progress in action theory depends on the integration of the philosophical with the scientific. Put another way, I advocate the naturalization of philosophical action theory.[8]

These are sentiments one can endorse, even cheer, but it is in proportion to the expectations they raise that Brand's book must be judged a failure, as I will try to indicate.

Brand's approach and outlook are best conveyed through following the structure of his book. Thus, a clue to the reasons for disappointment and exasperation can be seen from the fact that it is only in the seventh chapter, fully two thirds of the way into the book, that there is any hint of a concern with science or its possible bearing on philosophical theories of action. Until this point, Brand indulges without embarrassment in precisely the kind of arid scholasticism which he had promised to usher out. Worse still, the bulk of these discussions is only very tenuously related to the problems of action at all.

The first two chapters of the book are, admittedly, concerned with philosophical theories of action, and it is clear that this is where Brand is most at home – namely, in the nitpicking and hairsplitting of ordinary language and conceptual analysis. Typical is Brand's discussion of the problems of 'antecedential waywardness' and 'consequential waywardness', which are somewhat pretentious terms applied to cases in which something happens in ways not intended. Beyond this much, however, I confess to being unable to see what more needs to be said about such 'problems', and certainly why they might deserve attention as specially *philosophical* remains unclear. Thus, for example, Dan the sheriff is a lousy shot and kills a robber by a ricochet, rather than directly.[9] Brand's discussion of this and similar examples fails to reveal why this might be worthy of serious attention as raising deep problems of any sort, though conceivably, at worst, it might cause some slight terminological awkwardness if one contrived hard enough to manufacture it. Nevertheless, cowboy movies evidently manage to convey the subtleties of such stories without solving the profound philosophical problems alleged to inhere in them. Labelling such commonplaces as the 'problem of consequential waywardness' seems to be merely empty jargon.

Much of Brand's discussion in this vein is a recitation of truisms, which does not even have the virtue of being easy to read. We encounter here Brand's *penchant* for spurious formalization which plays no part his analysis and only serves to make the text unreadable. For example Brand writes:

> *If* (i) S has an action plan P such that his $A_i$ing and his A_{i+1}ing are contained in P but his A^*ing is not contained in P, (ii) S's A^*ing occurs between S's $A_i$ing and S's A_{i+1}ing or during S's $A_i$ing or S's A_{i+1}ing and (iii) S's A^*ing is not dependent on either S's $A_i$ing or S's A_{i+1}ing, or conversely, *then* S's $A_i$ing and S's A_{i+1}ing are intentional actions.[10]

Thankfully Brand translates this into a simple sentence of plain English, and he makes no further use of the formal statement.

Brand's obvious attachment in these first chapters to the old ways of doing philosophy of action is an early sign of what is to come and of his attitude towards the scientific material he examines. In this respect at least, Brand appears like the anatomist Vesalius who, while ushering in modern anatomy from its medieval origins, is nevertheless unable to rid himself of galenic principles, in spite of his own contrary evidence (though this is rather too flattering, since Vesalius, after all, succeeded).

Although Brand titles a section of his book 'The Ontological Foundations For Action Theory', this euphemistically disguises its very tenuous connection with actions. Thus, chapter three is a detailed discussion of the ontological status of events in general and a defence of Brand's favoured view of them as spatiotemporal particulars. Whatever may be the merits of the detailed discussion of such questions as the semantical interpretation of identity conditions for events, there is no attempt within the chapter or elsewhere to connect these considerations with the question of actions. Similarly, chapter four is concerned with the ontological status of the objects of propositional attitudes in general. Again, quite aside from a token gesture towards the bearing on action theory, naturalized or otherwise, there is no thread connecting the discussion with what has preceded or what is to follow. Plainly enough, Brand has utilized previously published papers (as he acknowledges in the preface), though without working them into a coherent, continuous essay. More importantly, their relevance to the problems of action is tenuous. Nevertheless, it is instructive to examine briefly some of the details of these discussions since they reveal a characteristic approach to the problem of action, and more generally, a characteristic attitude to the place of science in philosophy.

3. EVENTS AND OBJECTS

In a chapter on events, Brand argues for the view that they are "relevantly similar to physical objects, differing from physical objects in that they do not fully occupy the spatiotemporal region in which they occur".[11] However, it is far from clear that events can be coherently thought of as occupying spatial location at all, partial or otherwise. Does their partial occupancy of spatial regions vary by degrees, and if so, according to what principles? Further, why are they incapable of complete occupancy and what is the maximum extent of their possible occupancy? Brand's view seems to entail that objects differ from events as a matter of degree. Moreover, in view of the familiar dependence of event identification on the infinite possible descriptions, it is not at all clear that they can be uniquely characterised, much less accorded spatial location. To a degree, Brand himself acknowledges such difficulties with his question: "Where, for instance are we to locate Richard's long-distance telephone call to New York?" – a question to which Brand concedes there is not any completely nonarbitrary answer, but nevertheless insists somewhat lamely that "the lack of a full complement of epistemic and pragmatic principles for location does not vitiate the ontological point that events occupy spatiotemporal regions".[12] However, if the absence of relevant grounds supporting a view does not count against it, one would like to know what Brand thinks would do so. To be fair, Brand takes the problems for events to be parallel with those allegedly arising for objects in this regard insofar as he suggests that neither can objects be accorded precise spatial location. However, the parallel is less than compelling and the examples offered by Brand of such difficulty for physical objects appear to be quite spurious. Thus, for example, he asks: "does the region occupied by my house include the detached garage?" or "does the region occupied by the train include the empty spaces in the dining car?" I find it frankly difficult to see the philosophical difficulty alleged to reside in these cases where it seems perfectly clear that the 'problem' is entirely trivial and definitional. Certainly there is nothing comparable to the difficulty of even making sense of the notion that an event occupies some spatial region. Does Xanthippe's becoming widowed occur where she is, or in the cell where Socrates succumbs to the hemlock? At the very least the case for events occupying spatial regions requires support beyond merely assimilating their case to that of physical objects. After all, it is precisely their difference from physical objects in this regard which creates the philosophical puzzle. A fundamental question here is the issue of what principles or criteria one can deploy to solve the puzzles about events. Among the operative principles which may be inferred from Brand's discussion, though not defended explicitly, is the convenience afforded by a given view to the parsing of event sentences in extensional first order logic,[13] though he relies as much on intuition and commonsense.

However, given Brand's stated naturalistic ambitions, it is noteworthy that he fails to consider the possible relevance of adverting to whatever may be the implications of our best available physical theories, which undoubtedly have something to say on the matter. Instead, and probably as a result of this decidedly unnaturalistic approach, Brand is prepared to countenance such views as that only macroscopic things that we commonly sense are physical objects, whereas "molecules and atoms, not to mention subatomic particles, are not physical objects".[14] This claim offends as much against commonsense as against science, and Brand's failure on this score is an early indication of what emerges later as an eccentric and problematic view of the relations between philosophy and science – a matter to which I will return.

4. ATTITUDES, ASCRIPTIONS AND PSYCHOLOGY

Brand's concern (chapter four) with the propositional attitudes is, somewhat incidentally, more relevant to action through the inclusion of intending as one of the attitudes. Brand is principally concerned with the ontological status of the objects of these attitudes such as believing and desiring. He considers two general approaches to characterizing the differences between believing, intending and desiring: On one view which he labels 'object assimilationism', all the attitudes have the same objects, but are different kinds of attitudes towards these common objects; the second view, 'attitude assimilationism' takes the attitudes to be in some sense essentially the same, but having different objects. Details aside, the discussion of these attitudes seems seriously marred by Brand's egregious failure to distinguish a theory of thoughts from a theory of their ascription. As Devitt[15] has noted, it is common practice to use the term 'belief' when speaking of belief ascription, though the application of the terms *de dicto* and *de re* to thoughts taken literally is a mistake. Thus, differences in thought ascription do not entail differences in the thoughts ascribed. Taking a thought to be *de re*, as distinct from a thought ascription, if it is a relation between a person and some object involves a confusion since all singular thought tokens purport to refer and relate a person to an object. Nevertheless, Brand takes *de re* belief in just this way and, predictably, slides from the propositional attitudes properly speaking into discussing *de re* attitudes as some category of psychological theory.[16] He asserts that the analysis of *de re* attitudes "must not hypothesize psychologically impossible processes" as if there could be some question of empirical hypotheses at all. *De re* attitudes are thereby taken by Brand to implicate substantive theories concerning the phenomena of attention, perception and memory[17] and to function in the taxonomy of psychological states.[18] Brand says further that the analysis of *de re* attitudes must be compatible with psychological reality. The discussion here seems irredeemably confused, hardly illuminated by the introduction of the latter vexed issue, and would not appear to repay the effort of disentangling it all.

5. INTENDING AND DESIRING

Chapters five and six comprise part III of the book which is titled 'The Folk Psychology of Intending, Desiring and Believing'. Chapter five deals with desiring and begins by asserting that "the primary concern of this chapter is to delioeate the folk psychological or commonsensical, concept of desiring".[19] However, in spite of this pivotal role that folk psychology is supposed to play in Brand's enterprise, he proceeds immediately to engage in a philosophical debate concerning the character of desire and gives no further indication that his analysis has anything to do with folk conceptions. Rather, he addresses himself to a philosophical thesis advanced by Sellars and Castaneda concerning the nature of desire and, specifically, the claim that desiring is a species of intending. Brand undertakes a defence of the thesis that desiring is distinct from intending. This platitude could certainly qualify as folk psychology, but it is perfectly clear from Brand's advocacy of a 'philosophical' thesis as one side of a debate, that he is not merely articulating folk theory. Brand makes no appeal to folk conceptions, and offers characteristically philosophical arguments to defend a claim whose striking banality must, therefore, be attributed to the philosophical and not the folk theory. The claim that intending is distinct from desiring is surely one which would not be thought to require defending by the folk whose theories are allegedly being articulated. This is surely as primitive a datum as one could hope to find in folk psychology. On the other hand, Brand's disputations on behalf of this 'thesis' are arguably remote from any doctrines implicit in folk theory and, whatever may be their merits, it would be hard to find them mirrored in folk conceptions. In the end, Brand takes his philosophical arguments to have *established* the distinction which is supposed to be found in folk psychology.[20] This procedure can only be understood in the light of what seems to be Brand's general failure to distinguish technical philosophy from folk psychology. The same conflation will be seen later in his indifferent use of the two in stating what he takes to be the foundations of his inquiry.

Overlooking these methodological problems, we can briefly examine the substance of Brand's arguments concerning intending and desiring. Against Sellars and Castaneda, Brand argues that desiring cannot be assimilated to intending in view of the differences he notes between them.[21] Among these is the claim that it is possible for a normal person to have incompatible desires but it is not possible for him to have incompatible intentions. Brand asserts[22] that having incompatible desires is not abnormal, but there is serious difficulty in imagining a normal person who has conflicting intentions. Brand asks: "What would it be like to be in a mental state such that one, say, *both* intends to eat a second piece of pie and at the same time intends to refrain from doing so?".[23]

But it seems that there is no asymmetry in this regard between desiring and intending, since such contradictions are surely no more rational or normal in the case of desires. It seems just as incoherent to desire something and at the same time to desire its contrary as in the case of intending. The plausibility of Brand's case for the possibility of apparently contradictory desires rests on an implicit principle of charity according to which we can attribute different reasons for desiring something from the reasons for not desiring it. Brand's case rests on inconsistently applying the principle of charity in the case of apparently inconsistent desires and not in the case of inconsistent intentions. Thus, in the same way, a person could conceivably *intend* to go on a strict diet and at the same time intend to eat the second piece of pie, thereby seemingly both to intend eating it and at the same time to intend not doing so.

These platitudes might be seen as implicit doctrines of folk psychology, though Brand makes no attempt to defend construing them thus. Brand's discussion here and elsewhere suggests that, for him at least, philosophical discourse itself is constituted by such analysis. However, merely dealing in platitudes does not in itself constitute a concern with folk psychology. Even the articulation of the philosopher's intuitions on such questions is not quite the same as a self-conscious concern with folk theories, since, after all, philosophers are not quite ordinary folk in the relevant respects.

Brand's failure to distinguish systematic theory from folk psychology is further revealed in the model he proposes for the 'act sequence'.[24] The model takes the form of a diagram with boxes inter-connected by arrows. The boxes purport to represent events while the arrows represent causality. Incorporated in the model are the elements of belief, desire and intention, but in addition to these features, the diagram also includes representations of perceptual feedback, belief and desire matrices, physiological needs and natural laws, none of which plausibly qualifies as a notion available to ordinary folk. Even if we overlook this difficulty, however, the theoretical coherence of the model is unclear since the boxes include a miscellany of items such as natural laws, personality traits and belief matrices which do not appear to be events as Brand required. Since he fails to give any explanation of this diagram, one is left in the dark about other oddities such as double headed arrows and the question of how these are supposed to represent causal relations.

6. SCIENCE OF THE WILL

The essence of Brand's concern in the book is to account for what he takes to be possibly "*the* fundamental problem" in philosophical action theory, namely, "to adequately specify the nature of the proximate cause of action".[25] For Brand this fundamental problem about action concerns the nature of the mental antecedent such that only events of this type initiate action. He says: "I take it to be a deep insight of folk psychology that the initiating mental event has both cognitive and conative features. Without conative features, action could not begin".[26]

Thus, Brand explains: "My objective is to show how folk psychological action theoretic concepts can be transformed into scientific psychological ones".[27] In short, Brand is concerned to seek a scientific foundation for the will.

Although in principle Brand's search for the will has the considerable virtue of making it a scientific enterprise rather than a metaphysical one, this virtue is greatly diminished by his eccentric view of science. Brand begins encouragingly enough by claiming[28] that he will argue that philosophical and scientific theories are continuous. This is the now familiar view of Quine's 'epistemology naturalized', and Brand reassuringly emphasizes that his approach is in the spirit of "some recent developments in epistemology, the philosophy of mind, the philosophy of language and value theory",[29] citing Fodor and Dennett as exemplars. However, Brand's conception of this continuity has little in common with the tradition he claims to represent. Instead of seeing a single enterprise with shared substance, goals and principles, Brand takes science to be significantly distinct from philosophy and folk psychology (the latter two not being clearly distinguished, as we have already seen). For Brand, folk psychology has a privileged status whose all but infallible dictates are to serve as the Cartesian foundation upon which to construct science. He explains "my goal is to build from the bottom up"[30] and "philosophical action theory is the foundation on which I build".[31] Elsewhere he asserts "philosophical considerations (. . .) provide constraints on psychological theorizing".[32] A further methodological principle enjoins us to "point directions on empirical issues" from the allegedly more secure foundation of philosophy and commonsense, and it is clear that Brand does not see himself aboard the Quine-Neurath boat. Brand's view that there is such a privileged vantage point outside science is justified on the grounds of the corrigibility of science,[33] which is again curiously Cartesian for a contemporary naturalistic philosopher: "Since the models used in these disciplines are evolving and, sometimes, changing radically, it would be a mistake to inexorably tie an account of action to specific models".[34]

In the same way, many advocates of pseudo-science profess such an agnosticism about our best theories, on the grounds of their corrigibility. With Brand, they make the spurious assumption that there is some alternative which is immune from revision. Accordingly, Brand disavows any concern to engage in actual scientific speculation or adjudicate between competing theories, which he says would not be appropriate since "this is not an essay in empirical psychology".[35] However, like Descartes before him, Brand not only predictably casts "sidelong glances" towards science, but indulges freely in substantive criticism and speculation in spite of his disavowals.

For Brand, 'continuity' between science and philosophy has a special and peculiar meaning, involving some kind of metamorphosis in which folk psychological concepts become scientific ones through what he calls 'progressive transformation'. In this way, for Brand,[36] philosophical considerations provide constraints on psychological theorizing and he speaks of the need for scientific psychology to be "tempered by folk psychology".[37] Accordingly, when Brand does consider certain psychological models or theories it is with the express intention of co-opting them to philosophical purposes for which they were, of course, never intended. Brand insists that it is "an essential ingredient in the folk psychological picture of action that it is initiated by a mental event".[38] We saw him declare it to be a "deep insight" that an action is initiated by "conative features", though this insight amounts only to his somewhat tautologous insistence that a person must be "moved to act" by an "immediate intention", and it is this insight which is deployed to assess the adequacy of systematic scientific theories. Furthermore, it is on such an *a priori* basis that he complains against scientific theories that they are inadequate when they fail to accommodate these conative features – just as medieval anatomists would blame a cadaver when it failed to confirm galenic principles. In his words, such theories appear to be "unlikely candidates for the basis of a transformation of the conative feature of immediate intention".[39] In a typical example, Brand considers Hullian Drive Theory and concludes that it is unsatisfactory through its failure to reflect distinctions found in folk psychology.[40] Brand explains:

> An essential ingredient in the folk psychological picture of action is that it is initiated by a mental event; but neither Hull's nor Spence's neobehaviourism captures this part of the folk psychological picture. In most general terms the drive and incentive constructs leave out the mental life of the subject in action.[41]

> Folk psychology includes a distinction between intentional and nonintentional action (. . .) But Hullian Drive Theory does not reflect this facet of the folk psychological picture.[42]

> (. . .) Drive Theory is unsatisfactory. No progress can be made in clarifying and articulating the conative feature of intention in terms of drive, since that construct fails to serve a scientific explanatory role. It is as if an articulation of the commonsense notions of force and motion were attempted to be gained by transforming them into constructs within Aristotelian physical theory.[43]

This latter analogy is bizarre, though symptomatic, since it rests on the wisdom of hindsight in correctly noting that we would not employ a theory known to be wrong. However, the relevant point is, surely, that we cannot do better than endorse our *best* theories. *Their* failure to accommodate some notion is ordinarily taken to be grounds for rejecting that notion. Of course, a more appropriate analogy to capture Brand's project is that of trying to 'progressively transform' the idea of caloric into the kinetic theory, or that of phlogiston into modern chemistry. The failure of this 'transformation' counts against caloric or phlogiston and not the modern theory. It is a telling irony that Brand's entire book with repeated analyses of the above kind can be read in just this way as making a strong case *against* the scientific respectability of conative features of action. On Brand's own account, just like caloric, conation fails to become 'transformed' into any plausible theory. Thus, Brand seeks to superimpose a philosophical map on a foreign terrain and concludes from the mismatch that the terrain is at fault. Or, like the English speaking tourist in Europe, Brand complains of all the bloody foreigners. Less figuratively, Brand only grudgingly concedes[44] that folk psychology might in principle need to be abandoned in the light of systematic scientific research. His unexplained commitment to the sanctity of our folk theories, it must be noted, is quite different from the mere confidence in their overall reliability and their secure place in a mature cognitive science to be seen in Fodor, for example. Instead, of some current scientific work Brand says:

> But these theories, no matter their appeal, do not fully explain human action. They do not provide a transformation of the folk psychological concept of action because they leave out its essential property of being caused by a conative event. (. . .) scientific psychology is not free to develop any arbitrary conceptual scheme; it is constrained by the conceptual base of folk psychology.[45]

> Action theory is naturalized by bringing its conceptual basis closer to that of scientific psychological theories, not by eliminating it in favor of some particular scientific psychological theory.[46]

To be sure, there are plenty of cases in which folk concepts end up finding a place in systematic science, but this is more through good fortune than any methodological imperative.

7. NEUROSCIENCE AND NAIVETE

Among Brand's choice of theories, old and new, onto which he might graft conative features, noteworthy for its glaring ommission is the vast literature of the neurosciences where we have, as it happens, the richest and most detailed understanding of motor control, including, of course, the causal antecedents of action. These cannot be discounted as irrelevant through the 'functionalist stratagem' due to their status as mere physical realizations. In the first place, neurological study proceeds simultaneously at the functional and realization levels, since the lower levels of anatomical and physiological investigation are frequently driven by, and illuminate, the problems at a functional level. Thus, for example, the celebrated work of Hubel and Wiesel on stimulus feature detectors involved the responses of individual nerve cells, but made a most significant contribution to the understanding of the functional properties of the visual system. Geschwind has recently made essentially this point, suggesting that:

> Some behaviours can probably be understood poorly or not at all if the neural substrate is not considered. (. . .) In other words, there are cognitive systems the understanding of which may depend on knowledge of the specific pattern of realization in the 'hardware' of the nervous system.[47]

In any case, even the complete autonomy of a top-down functional analysis would be no justification for actually discarding insights from the bottom-up, particularly when we have, as in this case, a wealth of detailed understanding (possibly because in the case of action we are dealing with peripheral rather than central systems[48]). The extensive neurological findings concerning motor control are not only pertinent here, but more secure and less speculative than the theories considered by Brand, though, as I have argued elsewhere,[49] they provide no better basis for making scientific sense of the notions discussed in much philosophy of action.

The same general point has been made most recently by Patricia Churchland reflecting on the experience of a philosopher who becomes acquainted for the first time with the standard cases of a neurological clinic:

> (. . .) each Thursday brought a fresh assault on the commonsense framework within which we standardly comprehend the behaviour of normal brains – the framework within which we think about awareness, reasoning, unity of self, connectedness of cognitive capacities, connectedness of concepts and beliefs and so forth. So long as the brain functions normally, the inadequacies of the commonsense framework can be hidden from view, but with a damaged brain the inadequacies of theory are unmasked.
>
> Again and again one is dumbstruck by the problem of how to make sense of what one observes in the neurology ward. Such data seem to me crucially relevant in coming to understand how deficient folk psychology really is and how little we know of the deeper capacities underlying the known cognitive capacities of the brain.[50]

The bizarreness of many clinical syndromes dramatically brings home the utter inadequacy of naive philosophical reflection. The normal phenomena of thought and action can become decomposed and dissociated in ways which could not possibly have been anticipated by a commonsense uninformed by neurological knowledge. Nevertheless, philosophers continue to base their theories of action on considerations derived from a naive layman's understanding which thereby provides not only the data but also constraints on philosophical theory as Brand explicitly requires. It is in this sense that Freud's comment on the symptoms of hysteria seem so apt to describe philosophical theories as well: like hysterical symptoms, the philosophical theories behave as if they knew no physiology or as if it did not exist. This is the case, moreover, for essentially similar reasons, namely, their being founded on commonsense conceptions rather than the utterly unpredictable facts of anatomy and physiology. In neglecting neuroscience on the one hand, and elevating folk psychology to a pre-eminent status on the other, Brand is only articulating the precepts which have been implicit and standardly followed in philosophical speculation about action.

It is instructive to examine a typical example where we see David Armstrong[51] found his philosophical theory of intentional action on considerations derived from an imagined case of the sudden, unexpected paralysis of a limb. Armstrong defends the notion of trying as an essential element in intentional action, and his argument turns on imagining what would happen in the course of an ordinary intentional action if the agent were suddenly "struck by an instantaneous and complete paralysis of the arm that is to move".[52] For the point I wish to make here, the details are not relevant, and Armstrong concludes that "it would seem true to say"[53] that, although the motion of the arm has been removed, a mental event of trying must be left. Thus, if the agent was trying in the imagined case, "it is natural to suppose"[54] that such a mental event of trying is present in *all* intentional action.

There is much that could be said about this argument. Firstly, the conclusion is surely a non-sequitur, since whatever the phenomenon of trying amounts to in the imagined case of unexpected paralysis, it could just as plausibly be itself a product or artefact of the circumstances suggested. That is, the very occurrence of the paralysis *produces* the apparent phenomenon of trying. Certainly, to the extent that phenomenology is relied upon here,[55] the victim's own report is consistent with this suggestion, since ordinarily we would not describe ourselves as trying to move in the same way as when thwarted by the paralysis. Armstrong considers this point,[56] taking it to be essentially a linguistic one about "what we naturally say", and attempts to explain it away with a competing intuition about what "we can say truly and unmisleadingly".[57]

Thus, even in such cases where there is no apparent conscious effort, Armstrong suggests "there seems to be no infelicity in using the word 'trying'".[58] Armstrong is surely correct as far as the question of linguistic usage is concerned, but the issue I wish to raise is precisely what is the interest of such conclusions regarding linguistic usage. To the extent that we are concerned with the underlying phenomena of actions, the postulation of a mental event of trying is surely little more than a tautologous way of describing the intentional action itself. It seems close to being vacuous to suggest, as Armstrong does, that tryings are mental events which are physical processes in the brain causing the action. This is surely little more than baptising the antecedents of intentional action, whatever these might be, and can hardly count as providing any content or substance to our understanding of the matter. Armstrong is certainly correct to ask: "But who has shown that (. . .) independent characterizations of tryings and attemptings are impossible?"[59] just as we frequently describe a cause in terms of its effect. Thus, Armstrong points by analogy to such cases in which we say, for example, that *parents* are the causes of their children. Armstrong is surely right in suggesting that no-one has shown them impossible, for, on the contrary, such independent characterisations are only too easy to propose. That is, such characterisations run the risk of being like the notorious *virtus dormitiva* explanations: tryings are causes of actions in just the same sense as *virtus dormitiva* is the cause of sleep-inducing effects. Or like Skinner's vacuous identification of a stimulus only after knowing the response,[60] Armstrong proposes merely that "there is no reason why many tryings should not be mental happenings 'arising in the soul from unknown causes'".[61]

My principal objection to Armstrong's discussion of this matter, however, is of a different character altogether. Underlying the foregoing criticisms is a deeper point about the whole approach we see adopted by Armstrong. The objection is not so much to the use of such cases as the imagined paralysis to inform our theories; rather it is to the failure to be consistent and thorough with regard to the use of such examples. It is clear that reliance on examples available to the layman's commonsense is an utterly arbitrary restriction on the range of facts taken into account. The obvious question is: Why restrict our data and source of puzzles to a corpus arbitrarily delimited by the relative ignorance of the dilettante, layman philosopher? As Patricia Churchland observes, "what can or cannot be imagined about the empirical world is not independent of what is already understood and believed about the empirical world".[62] Moreover, it is important to appreciate that, however plausible this point, it is not merely abstract advice in principle, but founded upon what is already overwhelmingly clear in practice from the slightest acquaintance with the neurosciences. That is to say, beyond the fact that there can be no principled reason for restricting philosophical attention to what the layman knows and imagines, there is already available a vast and intriguing range of data and puzzles pertinent to understanding human action. Moreover, much of this material throws into disarray our commonsense notions of acting and intending – notions upon which philosophical theory has been based. A brief example will perhaps serve to give the flavour.

The apraxias are disorders of skilled, learned movement which are defined by exclusion, as not being accounted for by weakness, incoordination, deafferentation (sensory loss), abnormal tone or posture, intellectual deterioration, incomprehension, inattention, or uncooperativeness.[63] Significantly in the light of my foregoing remarks, Geschwind observes: "This definition may seem surprising, and one might even question whether such disorders could exist".[64] Like many clinical neurological syndromes, the apraxias present utterly paradoxical phenomena which could neither have been anticipated or imagined, nor can they be reconciled with our commonsense understanding. Thus, Geschwind explains, an apraxic patient might carry out commands with his right arm such as combing his hair, using a hammer or waving goodbye. But when given the same commands for his left arm, the patient would either make no response or make a clearly incorrect response. However, the patient could use his left arm quite normally for these very tasks under other circumstances. For instance, the patient could imitate any of these actions with his left arm when performed by the examiner, or he could use a hammer or comb normally when handed these objects. It is clear that the failure to respond to the verbal command could not be due to weakness, incoordination or any other elementary disorder of the left arm since it exhibited fully correct performances under other conditions.

Furthermore, the correct response of the patient with his right arm rules out inattention, incomprehension or uncooperativeness. The problem for the neurologist is how to account for these paradoxical results, and there is, in fact, an explanation in terms of specific lesions from which these observations can be seen to follow as inevitable consequences. However, my interest here is not so much in the neurological details of this problem, but rather in the very *prima facie* paradoxicality of the phenomena and the manner in which they illuminate the pointlessness of neurologically naive philosophical speculation. On the face of it, such apraxias appear to present cases in which intentional actions can be performed with one hand upon command but not with the other, even though there is nothing wrong with the hand in question. Equally bizarre manifestations of apraxia include the inability to demonstrate how to blow out a lighted match or suck through a straw by mime in the absence of the objects, but to perform normally when given the objects. Again, an apraxic patient might perform dance steps on command, or properly assume the position of a boxer when asked, but having adopted this posture and being asked to punch, the patient is unable to respond. Of these latter cases, Geschwind explains:

> Clearly the class of responses on command which is preserved is not distinguishable on linguistic grounds from those which are not carried out. Why should responses to 'Look up', 'Close your eyes', 'Walk', 'Bow', be preserved, while 'Make a fist', 'Salute', 'Wave' are lost? Simplicity is not an adequate explanation; three part commands such as 'Stand up, turn around twice, and then sit down' and 'Show me the position of a boxer' hardly seem more simple that 'Salute' and 'Make a fist'.[65]

Beyond noting again and emphasizing the fact that there are perfectly satisfactory neurological explanations of these phenomena, I will refrain from indicating what these are. This will serve to maintain the acute sense of paradox and thereby bring into relief the inadequacy and pointlessness of naive speculation of the kind we have seen practiced by Armstrong and promoted by Brand.

Of course, this example provides essentially the same lesson to be learned from the much discussed 'split-brain' work of Sperry and Gazzaniga. The interest of these commissurotomy cases derives not from their being intrinsically puzzling or inexplicable, for they are, on the contrary, precisely what one would expect given the damage to the 'wiring' involved. The moral to be drawn from these cases is only that they are bizarre and unexpected given our prior expectations derived from naive folk psychology. That is, the puzzle is not so much one of neurology as naiveté. This lesson can be multiplied indefinitely from the cases of the neurological clinic. Dejerine's syndrome (alexia without agraphia) for example, is a condition in which the patient cannot read even though there is nothing wrong with his eyesight; he can, for example, copy the letters and words of any text but is unable to understand it; nevertheless, he retains the ability to *write*, and may write a letter without difficulty, but is then unable to read what he has himself written! Such examples appear surprising and bizarre only because we expect the abilities of reading and writing to be somehow inextricably linked – even 'conceptually' linked perhaps. So much for the value of conceptual analysis and 'what one would say' in these cases. As Churchland notes wryly of similar examples, philosophers have even been known to deny the very possibility of dissociations of this kind.[66] Like Freud's hysterics, such philosophers have conceptualized the phenomena of mind and action in terms of their commonsense ideas which may have little or no relation to the underlying neurological reality.

Much of the recent debate over folk psychology has been an attempt to divine the ultimate fate of its concepts and theories in a mature cognitive science and, in particular, its disposition in relation to a completed neuroscience. Now, it seems to me that much of this philosophical fortune telling is inevitably a futile pursuit and, indeed, I am not at all sure why folk psychology holds any intrinsic interest at all, except perhaps to anthropologists. On the other hand, there is perhaps a reason, after all, quite different from the usual one, for attending closely to the substance of folk psychology, at least in the domain of action theory. Though this interest is rather different from whatever motivates the current fashion, folk psychology can take on a crucial heuristic value if, as I believe is the case, it is the source of deep philosophical muddles and conceptual confusions. In fact, I believe that much of philosophical concern with the alleged problems of action derives precisely from such sources in folk conceptions. I have tried to defend the view that these folk conceptions are, indeed, a species of deep cognitive illusion about ourselves – a kind of inherent pathology. Folk psychology derives such interest as it has only from providing the symptoms of a pathology we should like to cure in our systematic cognitive science.

Now, I believe that these assertions are not the kind of fortune telling which I have just deplored since the neurophysiological understanding of motor processes is already relatively rich, detailed and well developed, and permits some preliminary views about the mapping of folk notions. Returning to Brand now, we see that, while remarkably ignoring this vast body of systematic science concerning the proximate causal antecedents of action, he offers characteristically naive speculations about actions being the effects of special mental events of 'immediate intention', being 'moved to act' by 'conative features' and similar, more or less vacuous or platitudinous explanations. Brand's commitment to folk psychology as providing *a priori* constraints on empirical theory seems to be clearly a wrong-headed strategy in principle, but when the available systematic science already suggests the pointlessness of folk concepts and theories, this is akin to advocating a flat earth: logically possible, but unconvincing.

8. WILLIAM JAMES AND IDEO-MOTOR THEORY

Brand's approach to the ideo-motor theory of William James amply illustrates the point I have been making. The ideo-motor theory is enjoying a revival, particularly since the demise of behaviourism has permitted taking ideas or internal representations seriously again. More specifically however, the detailed substance of the theory provides an elegant and plausible account of some of the complex features of motor control. For example the ideo-motor theory can accommodate the inadequacies of the serial chaining and closed loop models, both of which rely on servo control mediated by reafferent information – correct responses in the former and incorrect responses in the latter. By contrast, the ideo-motor theory purports to explain precisely the *initiation* of an action through an idea or anticipatory image of its own sensory feedback response. In the closed loop model, such an image may serve as a template for comparison with current feedback information, but it need not be activated prior to the action. On the other hand, the ideo-motor model postulates an image as serving actively prior to the performance in initiating movement.[67]

Details aside, and whatever its merits, it is clear that James's ideo-motor theory puports to be an exhaustive account of the necessary and sufficient conditions of voluntary action[68] – that is, precisely the "proximate causes" of action whose discovery Brand takes to be the "fundamental problem". James's theory obviates the necessity for postulating mysterious supernumerary volitions in an eliminative sense. It is simply question-begging to complain as Brand does that this account leaves out the 'conative features' of action or the reason why one is 'moved to act'.

It is interesting to look a little more closely at James's writing on this subject, for he appears to speak clearly and directly to Brand where he writes:

> The question is this: Is the bare idea of a movement's sensible effects its sufficient mental cue, or must there be an additional mental antecedent, in the shape of a fiat, decision, consent, volitional mandate, or other synonymous phenomenon of consciousness, before the movement can follow?[69]

Taking up precisely the issue we saw raised by Armstrong too, James suggests that "the bare idea is sufficient" in the normal cases of intentional action, and it is only when there is some "antagonistic representation present simultaneously to the mind" that an "express fiat" or volitional act is required.

We may then lay it down for certain that every representation of a movement awakens in some degree the actual movement which is its object; and awakens it in a maximum degree whenever it is not kept from so doing by an antagonistic representation present simultaneously to the mind.

The express fiat, or act of mental consent to the movement, comes in when the neutralization of the antagonistic and inhibitory idea is required. But that there is no express fiat needed when the conditions are simple, the reader ought now to be convinced.[70]

Then, continuing, James directly addresses Brand's insistence on the need for 'conative features':

Lest, however, he should still share the common prejudice that voluntary action without 'exertion of will power' is Hamlet with the prince's part left out, I will make a few farther remarks. The first point to start from in understanding voluntary action, and the possible occurrence of it with no fiat or express resolve, is the fact that consciousness is in its very nature impulsive. We do not have a sensation or a thought and then have to add something dynamic to it to get a movement. (. . .) The popular notion that mere consciousness as such is not essentially a forerunner of activity, that the latter must result from some superadded 'will-force', is a very natural inference from those special cases in which we think of an act for an indefinite length of time without the action taking place. These cases, however, are not the norm; they are cases of inhibition by antagonistic thoughts. When the blocking is released we feel as if an inward spring were let loose, and this is the additional impulse or fiat upon which the act effectively succeeds. (. . .) But where there is no blocking, there is naturally no hiatus between the thought-process and the motor discharge.[71]

At the very least, we can safely say that the discussion does not appear to have advanced greatly since James's time, for we see here a striking answer to precisely the earlier concerns of Armstrong.

In passing, it is perhaps worth making a further observation on Brand's treatment of the James theory. Brand characterizes this as a "mechanical model"[72] and as "a theory about the physical systems of persons and other animate beings" which he contrasts with "another approach to understanding guidance and monitoring that is more abstract". This other approach, Brand explains, does not focus on particular mechanisms, but relies on the analogy to computers in delineating a structure or functional organisation. Brand explicitly takes this distinction to correspond to the hardware/software distinction and thereby clearly intends to place ideo-motor theory in the former category. However, this is utterly misconceived since there is nothing at all in the theory which commits it to any particular physical substrate or realization, and it is, on the contrary, a paradigm case of an abstract, functional, information processing model.

9. 'TOTE' MODEL FOR ACTION

With the move toward naturalization, the task of the philosopher as amateur and dilettante in the highly specialized cognitive sciences has become admittedly a difficult one, as Dennett has candidly avowed.[73] Specifically in the philosophy of action, Danto has significantly remarked: "It is perhaps impossible to overemphasize (*mea culpa!*) the degree to which a certain physiological recklessness has dominated the philosophy of action in recent times".[74]

Brand actually wants to turn this vice into a virtue by making the neglect of science into a principled stance, pretending to eschew scientific speculations, while, for all his protestations, he openly engages in substantive theorising, such as it is. Thus he offers a critique of various theories such as that of Miller *et al.*, and it is instructive to look briefly at Brand's treatment of this particular theory both because of the theory's central importance and also because of the light it sheds on Brand's approach.

Brand begins by giving a confused and misleading statement of the status and significance of the classic TOTE schema of Miller, Galanter and Pribram saying: "The arrows represent the flow of information. In a physical system that instantiates this unit the arrows depict the transfer of energy".[75]

Firstly, in this brief and inadequate statement, Brand seems to imply that the flow of energy is somehow to be construed as the physical realization of the flow of information. However, independently of the TOTE model, this is quite confused since energy and information function as independent theoretical constructs and alternative kinds of analyses, much as one might measure the voltage or listen to the conversation in a telephone line. Quite apart from this, however, Miller *et al.* point out that the arrows are open to interpretation in *three* quite different ways according to which they can represent either energy, or information, or the flow of control, depending on the level of abstraction chosen.[76] Indeed, *most* important for their account is this third conception in which it is the distinct and "intangible something called *control*" which flows over the arrows of their familiar diagram, though remarkably Brand makes no mention of this at all.

Brand proceeds to make an obscure and dubious distinction between static and dynamic theories, though he seems to have confused *theories* with the *phenomena* they are intended to explain. Be that as it may, unaccountably, Brand takes the Miller, Galanter and Pribram model to assume that subjects are constantly active, constantly executing some plan or other, though such an assumption is surely independent and irrelevant to the schema they give for actions. Brand takes this to be a failing of "dynamic theories", which he says "misdescribe human action".[77] He also concludes that the TOTE model is not "fully scientific"[78] and, furthermore, it is also "viciously regressive".[79] These are decidedly odd conclusions to arrive at (especially for one who has claimed to eschew substantive scientific theorising), in view of the status of the TOTE theory. Whatever may be its shortcomings, it has come to be the standard analysis of human action in both top-down and bottom-up theories,[80] for it is after all, little more than an abstract schema characterising the central role of feedback or servo-mechanism control. In this sense it is a highly successful explanatory model and, in any case, the best we have got. It is characteristic of Brand throughout, and as we will see, a tacit principle, that he should fail to give such theories their due, which is, of course, not to say that they are beyond criticism. However, such severe criticisms of what is a well-entrenched and successful explanatory model would surely deserve more than more than Brand's critical offerings to warrant the extreme conclusions he draws.

Thus, for example, introducing his own locution, Brand speaks of a theory being "depth scientifically functional (. . .) if it decomposes complex events into primitive operations and gives a functional account of these operations by means of a scientific psychological theory".[81]

In the first place, according to Brand, the TOTE model fails to be 'fully scientific' because some of its key notions such as 'plan' and 'value' are taken from folk psychology. He complains: "'Plan' and 'value' are taken from folk psychology, and while there is some attempt to embed them in scientific psychological theory, that attempt is far from complete or successful".[82]

It is difficult to see why this should be counted against the TOTE model, especially coming from Brand, since it seems to be precisely the kind of 'transformation' he aspires to for his own folk psychological notion of conation. Nevertheless, on this view physics would have to count as not fully scientific as well, to the extent that it has co-opted such everyday terms as mass and heat. Of course, the point is that such terms take on a different and precise technical meaning in the context of the theory in which they come to find a role. Far from being a failing, it is a sign of progress.

Brand observes further that the notion of the TOTE unit itself is not folk psychological, "though it suffers from not receiving precise and systematic development".[83] This would appear to be an entirely misguided remark concerning what is, after all, the paradigm of systematic scientific development in the field. It is a symptom of Brand's highly eccentric views of science and its relation to folk theory, of which we will see further manifestations presently.

Having already claimed to have demonstrated a vicious infinite regress, among other fatal flaws, it is difficult to grasp how Brand could see any problems being even more important, and thus in proposing what he takes to be 'the most important' problem, a sense of attempted overkill hints at the lameness in each of the arguments. Nevertheless and perhaps predictably, Brand's principal complaint against the TOTE model is that it is "purely cognitive, and does not capture the conative feature of action".[84] Brand never considers the possibility that such theories as that of Miller *et al.* should be construed as exhaustive accounts and, in that sense, eliminative in a familiar philosophical sense.

10. CONCLUSION: CONATION AS '*WILL O' THE WISP*

Whatever may be the difficulties and obscurities of Brand's practice, we might turn to his explicit methodological principles in the hope of clarifying the problems I have indicated. However, what Brand offers here only serves to confirm the diagnoses I have made. Tucked away inconspicuously as an unlikely subsection of chapter six on intending and believing, Brand offers an account of the relation of folk psychology to systematic science. He begins by stating that both folk psychology and scientific psychology can be modelled by the formalism of ordered pairs. Folk psychology can be represented as $<K, G>$ where K is a set of kind terms and G a set of generalizations in which the members of K occur essentially; scientific psychology, according to Brand, has the structure $<N, L, T>$ where N is a set of natural kind terms, L a set of law-like generalizations in which the members of N occur essentially, and T is a set of theories in which the members of L occur essentially. Whatever might be the point of this formalism, it is utterly flatulent here since it is unexplained and unilluminating, playing no further role in the exposition at all.

Brand proceeds to introduce a notion of 'conceptual psychology'[85] which he takes to be certain claims shared by folk psychology and scientific psychology additional to their empirical claims:

(. . .) on the whole there is a recognizable tripartite distinction between folk psychological generalizations, scientific psychological generalizations, and conceptual claims that support and define both types of generalizations.[86]

In case this is not transparently clear, Brand explains 'conceptual psychology' by saying that it can be modelled by yet another formalism:

Conceptual psychology, the common base of scientific and folk psychology, can itself be modeled by a three-dimensional space, where points in the space are statements. Let us specify that the x-coordinate is the degree of theoretical relevance, the y-coordinate the degree of empirical import, and the z-coordinate time.[87]

In case this is still obscure, Brand offers examples of 'conceptual psychology' such as the following: 'there is a linguistic code in which persons think', 'mental events have syntactical (formal) properties',[88] 'some thinking is imagistic'.[89] However, even if one didn't have reservations nowadays about distinguishing analytic from synthetic as Brand wishes to with his category of 'conceptual psychology', these statements would appear to be straightforward paradigms of empirical, theoretical claims. By this time one might justly conclude that any unclarity and confusion is Brand's fault.

Brand's method recommends "the change of conceptual claims underlying folk psychology into ones that underlie scientific psychology",[90] calling this process 'progressive transformation', which in an idealized form he labels 'the Method of Case Studies'. As we have seen, in essence this method involves taking some folk psychological notion, finding some pertinent theory such as the TOTE model of Miller *et al.*, or William James's ideo-motor theory, and finally: "Transform[ing] the original, unacceptable conceptual claim into one that has higher x- and y- values".[91]

To the extent that any of this is clear, it seems safe to say that, construed descriptively, it is a dubious account of scientific practice, and construed normatively, the obvious question is: Why bother? Without a galenist's prior convictions concerning the *rete mirabile* this elaborate and contrived algorithm for scientific progress has little to recommend it.

Since the issues surrounding the role and status of folk psychology are central to Brand's project, and particularly since he holds somewhat unorthodox views on the matter, it is a curious shortcoming of Brand's book that he nowhere makes reference to the recent important discussions of just this question by philosophers such as Churchland and Stich. The absence of Stich's important book from the bibliography may be an oversight.

Brand's book ends abruptly, seemingly in the middle of an inconclusive discussion of recent theories of intentional action. Turning back a few pages, however, the reader finds remarks which were made in passing, almost offhandedly, but which are, in fact, Brand's central thesis and final verdict. As in the case of every other theory examined, Brand says that his complaint against the one under discussion "is that the noncognitive features of action initiation have been for the most part ignored".[92]

Thus, in the end, even on his own account, Brand's project is unsuccessful as he laments: "My original negative thesis remains: there is presently no extant theory or (or theories) adequate to transform fully the folk psychological concept of conation into a scientific one".[93]

Indeed, Brand had made this confession earlier in the book, but as the ever hopeful father trying to find a respectable suitor for the uglier of his two daughters, he remains optimistic:

> Unfortunately the situation for the conative features is less happy [than for the cognitive ones]. There is presently no tenable scientific theory into which this folk psychological concept can be transformed. (. . .) The best that can be currently achieved is a partial transformation. A complete transformation of conation must await future scientific developments. Only informed speculation is now possible.[94]

The implication of this last remark is that 'informed speculation' can somehow go beyond what our best science itself can warrant. Of course, on the contrary, our scientific theories *constitute* our best 'informed speculation' and here we see Brand using a stratagem popular with purveyors of the paranormal: he takes his failures to reflect adversely on science rather than on his favoured theory. Thus, in the absence of positive evidence, he resorts to the entirely vacuous claim that future science will vindicate him, but he thereby unwittingly places conation in the same category as all the other things which presently have nothing whatever to recommend them.

NOTES

[1] P.M. Churchland (1986) 'Some Reductive Strategies in Cognitive Neurobiology', *Mind* **XCV**, No. 379, pp. 279–309, and 'Cognitive Neurobiology: A Computational Hypothesis for Laminar Cortex', *Biology and Philosophy* **I**, pp. 25–51.

[2] P.S. Churchland (1986) *Neurophilosophy*, MIT Press, Cambridge, Mass.

[3] D.M. Clark (1982) *Descartes' Philosophy of Science*, Manchester University Press, Manchester.

[4] P. Slezak (1986) 'Actions, Cognition and the Self', *Synthese* **LXVI**, pp. 405–435.

[5] My argument here has been in the spirit of an interesting remark made by D.M. Armstrong in (1973) 'Epistemological Foundations for a Materialist Theory of the Mind', reprinted in (1980), *The Nature of Mind and Other Essays*, University of Queensland Press, pp. 32–54. Here he suggests that the plausibility of materialism would be considerably enhanced if an independent account could be given of why it seems introspectively *im*plausible. I have attempted to give just such an independent materialist account from which these very introspective intuitions follow as predictable consequences.

[6] G.A. Miller, E. Galanter and K. Pribram (1960) *Plans and the Structure of Behaviour*, Holt, Reinhart and Winston.

[7] M. Brand (1984) *Intending and Acting: Toward a Naturalized Action Theory*, Bradford /MIT, p. x.

[8] *Ibid.* p. x.

[9] *Ibid.* p. 18.

[10] *Ibid.* p. 29.

[11] *Ibid.* p. 51.

[12] *Ibid.* p. 56.

[13] *Ibid.* p. 53.

[14] *Ibid.* p. 58.

[15] M. Devitt (1984) 'Thoughts and their Ascription' in P.A. French, T.E. Ueling and H.K. Wettstein (eds.) *Midwest Studies in Philosophy*, University of Minnesota Press, pp. 385–420.

[16] See *op. cit.* (note 8), p. 107.

[17] *Ibid.* p. 114.

[18] *Ibid.* p. 115.

[19] *Ibid.* p. 121.

[20] *Ibid.* p. 126.

[21] *Ibid.* p. 125.

[22] *Ibid.* p. 126.

[23] *Ibid.*

[24] *Ibid.* p. 128.

[25] *Ibid.* p. 33.

[26] *Ibid.* p. 45.
[27] *Ibid.* p. 191.
[28] *Ibid.* p. x.
[29] *Ibid.*
[30] *Ibid.* p. xi.
[31] *Ibid.*
[32] *Ibid.* p. 46.
[33] *Ibid.* p. xi.
[34] *Ibid.*
[35] *Ibid.* p. 174.
[36] *Ibid.* p. 46.
[37] *Ibid.* p. 157.
[38] *Ibid.* p. 247.
[39] *Ibid.*
[40] *Ibid.* p. 248.
[41] *Ibid.* p. 247.
[42] *Ibid.* p. 248.
[43] *Ibid.* p. 249.
[44] *Ibid.* p. 168.
[45] *Ibid.* p. 239.
[46] *Ibid.* p. 169.
[47] N. Geschwind (1980) 'Neurological Knowledge and Complex Behaviours', *Cognitive Science* **IV**, No. 2, p. 192.
[48] J. Fodor (1983) *The Modularity of Mind*, Bradford.
[49] See *op. cit.* (note 4).
[50] See *op. cit.* (note 1), p. 223.
[51] See Armstrong, 'Acting and Trying', in *op. cit.* (note 5), pp. 68–88
[52] *Ibid.* p. 69.
[53] *Ibid.*
[54] *Ibid.*
[55] *Ibid.* p. 70.
[56] *Ibid.* p. 71.
[57] *Ibid.* p. 72.
[58] *Ibid.*
[59] *Ibid.* p. 75.
[60] N. Chomsky (1959) 'Review of B.F. Skinner, Verbal Behaviour', *Language* **XXXV**, pp. 26–58.
[61] See *op. cit.* (note 5), p. 75.
[62] See *op. cit.* (note 2), p. 3.
[63] N. Geschwind (1975) 'The Apraxias: Neural Mechanisms of Disorders of Learned Movement', *American Scientist* **LXIII**, pp. 188–195, and K.M. Heilman and L.J. Gonzalez Rothi (1985) 'Apraxia', *Clinical Neuropsychology*, K.M. Heilman and E. Valenstein (eds.), Oxford University Press.
[64] Geschwind, *op. cit.* (note 63), p. 188.
[65] *Ibid.* p. 192.

[66] See *op. cit.* (note 2), p. 223.
[67] A.G. Greenwald (1970) 'Sensory Feedback Mechanisms in Performance Control', *Psychological Review* **LXXVII**, pp. 73–99.
[68] W. James (1890) *The Principles of Psychology* Volume **II**, Henry Holt and Company, republished by Dover, 1950, p. 522 and p. 526.
[69] *Ibid.* p. 522.
[70] *Ibid.* p. 526.
[71] *Ibid.* p. 526–527.
[72] See *op. cit.* (note 7), p. 181.
[73] D. Dennett (1978) *Brainstorms*, Bradford, p. 109.
[74] A.C. Danto (1979) 'Basic Actions and Basic Concepts', *The Review of Metaphysics* **XXXII**, p. 474.
[75] See *op. cit.* (note 7), p. 182.
[76] See *op. cit.* (note 6), p. 27.
[77] See *op. cit.* (note 7), p. 184.
[78] *Ibid.* p. 185.
[79] *Ibid.* p. 186.
[80] N. Harvey and K. Greer (1980) 'Actions: The Mechanism of Motor Control', *Cognitive Psychology: New Directions*, G. Claxton (ed.), Routledge and Kegan Paul.
[81] See *op. cit.* (note 7), p. 182.
[82] *Ibid.* p. 185.
[83] *Ibid.*
[84] *Ibid.* p. 187.
[85] *Ibid.* p. 183.
[86] *Ibid.* p. 163.
[87] *Ibid.* p. 165.
[88] *Ibid.*
[89] *Ibid.* p. 163.
[90] *Ibid.* p. 167.
[91] *Ibid.* p. 168.
[92] *Ibid.* p. 256.
[93] *Ibid.* p. 263.
[94] *Ibid.* p. 237.

Centre for Cognitive Science
University of New South Wales, Sydney, Australia

PHILIP CAM

NOTES TOWARD A FACULTY THEORY OF COGNITIVE CONSCIOUSNESS

On the view to be expounded below, the coordination and control of a person's behaviour, insofar as it makes reference to cognition, is accomplished by a set of faculties. As I am going to construe things, a faculty is a storage and production facility which can be characterized by its representational modality or format, the kinds of operations it typically carries out over representations in that format, and its processing connections to transducer systems, effector systems, and other faculties, in the production of behaviour. In addition, and most centrally, I claim that consciousness is a faculty affair. Each faculty is responsible for producing conscious states within its representational and operational domain – indeed, consciousness just is the running result of operations in these faculties – and insofar as coordination between faculties is dependent upon these results it is dependent upon consciousness. So it is an error to deny consciousness a systematic role in cognition.

1. PRELIMINARY REMARKS

1. One preliminary indication of the sort of thing that I have in mind may be provided by way of comparison with the faculty theories put forward by Jerry Fodor and Howard Gardner.[1] First, according to Fodor, the human cognitive apparatus can be divided functionally into three basic types of systems: transducers, input systems, and central systems. Transducers map proximal stimuli onto covarying neural firing patterns; input systems respond to transducer outputs and perform at least partial perceptual analyses which they deliver up to central systems; and central systems mediate higher cognition or thought. Further, Fodor's input systems are highly modular processors. Principally, that is to say, they are *domain specific*, in that they respond to a relatively specific range of inputs, and produce a relatively restricted range of outputs; and they tend to be *informationally encapsulated*, in that their computational access to information derived from other stimulus domains and from higher cognition is quite limited. By contrast with the input systems, Fodor's central systems lack domain specificity and

Peter Slezak and W.R. Albury (eds.), Computers, Brains and Minds, 167–191

informational encapsulation. They are characteristically non-modular. Now, while I reject Fodor's holistic account of higher cognition,[2] I am inclined to think that his story about input systems is largely correct. And what Fodor says about input systems can be transferred, at least in part, to a thoroughgoing faculty theory. In particular, I shall argue that faculties are in important respects both domain specific and informationally encapsulated. However this should not be taken to imply that I wish to divide the systems mediating higher cognition, in their turn, into higher-order modules along the lines of Fodor's input systems. Rather, I am proposing a cognitive architecture on a larger scale than Fodor's input systems, and one which need not respect the boundary between input analysis and higher cognition. My claim is that there are significantly modular systems which are altogether more molar than Fodor's modules, and that these modular systems, or faculties, are responsible for both perception and cognition.

Howard Gardner's theory of multiple intelligences is somewhat closer to the current proposal. Gardner distinguishes between linguistic, musical, logico-mathematical, spatial, bodily-kinesthetic, and personal intelligences, on a range of criteria, including isolation by lesion, the kinds of identifiable operations an intelligence typically carries out, evolutionary considerations suggested by animal comparisons, experimental psychological results relating to such things as transfer of training and interference effects, and the different forms of public symbol systems, in which individuals have quite differential skills, and which may be taken to exploit different cognitive capacities. It seems to me, however, that these intelligences are broader cognitive capacities than those I am attempting to identify. While Gardner's intelligences, I would assume, find much of their basis in the collective capacities of different ranges of faculties, they are not to be individually identified with them. If we take spatial intelligence, for example, we find a collection of abilities associated with different domains of experience:

> Central to spatial intelligence are the capacities to perceive the visual world accurately, to perform transformations and modifications upon one's initial perceptions, and to be able to re-create aspects of one's visual experience, even in the absence of relevant physical stimuli. One can be asked to produce forms or simply to manipulate those that have been provided. These abilities are clearly not identical: an individual may be acute, say, in visual perception, while having little ability to draw, imagine, or transform an absent world.[3]

Clearly, visual perception, mental imagistic capacity, and kinesthetic or body sense are amongst those non-identical abilities which make important contributions to what Gardner calls spatial intelligence. They are also associated with different domains of experience. On my view, they are capacities of different faculties, which contribute to spatial intelligence. At the level of the present analysis, then, cognitive architecture incorporates a range of semi-autonomous processing systems more restricted than the collective arrangements which embody Gardner's intelligences.

2. Another way of bringing out the general scheme of things being proposed here is to consider a case where the normal patterns of activity of the cognitive system can selectively be made to break down. Let me begin with a study reported by Michael Gazzaniga and Joseph LeDoux.[4] In discussing engram transfer, Gazzaniga and LeDoux consider a sodium amytal test carried out on left-dominant subjects. The left hemisphere is temporarily put out of commission by injecting sodium amytal into the blood supply to that hemisphere. After the injection has taken effect, the subjects are given a commonplace object, a spoon say, to hold briefly in the left hand. When the left hemisphere is back in operation and the subjects are asked to say which object was presented, they typically deny knowledge of the stimulus. Yet the subjects have no difficulty in pointing with their left hand to the correct object on a chart depicting a range of commonplace implements. The explanation supplied for this disparate performance is as follows. Both motor control and sensory projections for the tactile perception of objects (stereognosis) for the left hand reside in the right hemisphere, while in left-dominant subjects speech control lies in the left hemisphere. On the one hand, then, we can say that the subject was able to select the correct item from the chart by drawing upon a memory trace formed in the right hemisphere. The lack of any verbalizable memory, on the other hand, is explained by suggesting that, while a memory trace derived from stimuli tactually presented to the left hand is normally also conferred upon the left hemisphere by callosal transfer from the right hemisphere, this transfer only occurs at the time that the subject perceives the stimulus; and since the left hemisphere was out of commission at the time, no transfer occurred. And given that speech control derives its information from left-brain storage, no trace was available for report. Now the point

that I wish to make about this interpretation of the study is that the subject's right hemispheric memory trace, which informs the subject's pointing response, is not accessible to the production systems for speech. And let us note that this is not an artefact of the experiment. The conclusion to draw is that retrospective verbal reports of objects presented to the left hand always draw upon left hemisphere memory traces. The verbal equipment simply has no direct access to the right hemisphere memory bank in which the memory of the tactile stimulus has been deposited. So here we have evidence of a right-brain storage facility which is accessible to systems generating behaviour through the left hand, but which cannot be accessed by the systems generating verbal behaviour.

I believe that we have here struck the tip of a very large iceberg. I want to suggest that there are a fair number of physically distinct storage and production facilities which are selectively linked, and that through their selective linkages these facilities form a cognitive architecture of coordinated faculties in the brain.

3. On a larger scale, the existence of physically distinct processing systems capable of a large measure of coordinated yet semi-autonomous functioning is well-known from the studies of split-brain patients.[5] Although serious behavioural deficits can be produced experimentally with these patients, I take it to be a matter of considerable significance that the physically discrete processing systems revealed by severing the corpus callosum are capable of subserving more-or-less normal psychological function in daily life even though these connections have been cut. The experimental deficits result from a combination of the physical isolation produced by severing the effective connections between the cerebral hemispheres and the carefully controlled lateralized procedures used in the tests. As I say, outside of the experimental testing conditions, these patients tend to exhibit roughly normal behaviour. The following comments from Sperry are typical:

> The first point to be emphasized is that these patients following surgery appear in ordinary, everyday behaviour to be very typical, single-minded individuals. (. . .) After recovery patients without other brain damage are able to return to school or to household duties, or to an undemanding job assignment. Two years after surgery, a typical commissurotomy patient without complicating disorders, could easily go through a complete routine medical examination without revealing to an uninformed practitioner that anything is abnormal. Nor is there any marked change in the verbal scores on the standard IQ

test. Complaints about short-term memory are common especially in the early years after surgery. However, the general behaviour and conversation during the course of casual social encounter without special tests typically reveals nothing to suggest that these people are not essentially the same persons that they were before the surgery with the same inner selves and personalities.[6]

This suggests that the integrity of behaviour, even in normal subjects, need not depend too heavily upon cooperation established through direct processing connections between the left and right brain systems. As I see it, there are three other contributing factors. First, there is what we may call the 'preestablished harmony' of the two systems. The sort of thing that I have in mind here is noted by Sperry when he says that "these two separated mental spheres have only one body, so they always get dragged to the same places, meet the same people, and see and do the same things all the time",[7] and by Puccetti when he adds, for instance, that the two systems share "common subroutines of learned behaviour stored in the still intact cerebellum; and a common internal milieu (blood sugar, hormone levels, etc.)".[8] In general, nature has so arranged the left and right hemisphere systems in relation to other psychological structures, and to a broad range of bodily conditions, that they are capable of acting as autonomous yet harmonious counterparts. Secondly, in addition to the direct routes of internal communication normally afforded by the corpus callosum, there are indirect links between these two systems through both other bodily routes and the environment. This allows for "cross-cuing" between the hemispheres.[9] Thirdly, there is the matter of cerebral dominance. As I understand it, cerebral dominance is no longer thought of as simply the control of complex behavioural functions by the so-called dominant hemisphere. Each hemisphere tends to develop its own specialized contributions to the behavioural repertoire. Indeed, this is precisely the conclusion which is widely drawn from the split-brain studies themselves. Here again is Sperry:

The mental performance of [the right 'non-dominant'] hemisphere after commisurotomy has been found repeatedly to be superior to that of the speaking hemisphere in a growing series of nonverbal, largely spatial tests. The tasks involved are of the kind where a single spatial image processed as a whole proves to be more effective than a detailed verbal or mathematical description. Examples include the copying of designs, reading faces, fitting forms into molds, discrimination and recall of nondescript tactual and visual forms, spatial transformations and transpositions, judging whole circle size from a small arc, grouping series of different sized and shaped blocks into categories, perceiving whole plane forms from a collection of parts and intuitive apprehension of geometrical properties. (. . .) Earlier doubts regarding the presence of advanced mental function in the minor hemisphere are now largely dispelled and the concept of a complementary evolution of both hemispheres has come to replace our older classic view of a single one-sided dominance.[10]

Cerebral dominance, then, suggests that each hemisphere has somewhat specialized systems of cognition which can effect behaviour concordant with their own expertise, and without relying on much in the way of direct internal communication between the system and major processing systems of the other hemisphere. Together these three factors caution against over-emphasizing the role of direct communications between these systems.

While it is not as yet clear to what extent the point that I have just been making in relation to the left and right hemispheres can be sustained for smaller processing systems, one of the suggestions on offer in this paper, at any rate, is that direct internal communication between faculties may be rather limited. This is not to say that internal communication is unimportant. For even if faculties do make essentially autonomous contributions to behaviour, and act in concert by other means, direct internal communication between faculties can be shown to play a vital role in the coordination of human cognition and attendant behaviour. Having entered my caveats, it is the latter means of coordination upon which I will focus.

2. FACULTY INTERACTION AND THE SPECIAL ROLE OF CONSCIOUSNESS

1. Let me now exhibit the range of features which I claim for faculties. Since I particularly want to bring out the role of consciousness in faculty interaction, I will begin from the phenomenological perspective. An example will help. Let us take Shepard and Metzler's widely discussed studies of mental imagery,[11] in which subjects are exposed to pairs of line drawings depicting either the same or somewhat different three-dimensional assemblies of cubes in different orientations (see Figure 1). In some cases the differences in orientation are merely in relation to the picture-plane (as in (A)) and in others they extend through three dimensions (as in (B)). In each case Shepard and Metzler's subjects are asked whether the two configurations are the same or different. Now the point of interest for us is that the subjects tend to report that they solve

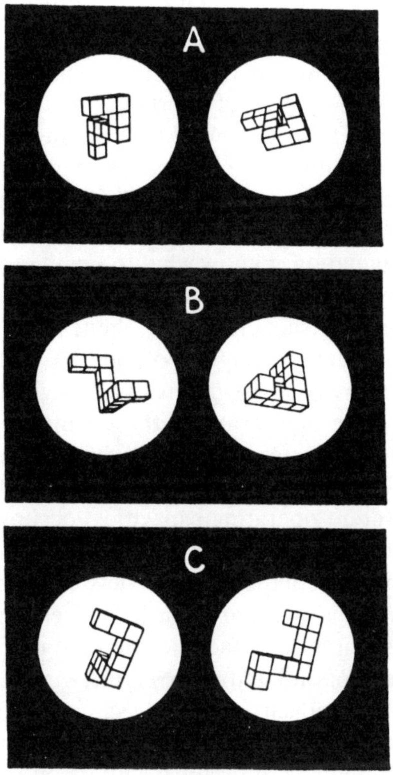

Fig. 1

this task by first consciously rotating one member of the pair to the perceived orientation of the other. Unfortunately I do not have their protocols, but it will suffice to reconstruct my own processing strategies. The aim in each case is to discover whether the two configurations are the same or different. With this in mind, I first briefly survey the two configurations, and get some initial impression of their structure and orientation. Then I survey the left-hand figure, attempting to analyse its configuration by running through the figure noting its twists and turns. When I feel that I have a fair grasp of the configuration I turn to the right-hand figure, looking to map the left-hand figure onto it. When I do this, in imagination I seem to swing the left-hand figure to the orientation of the right-hand figure. I perceive that it fits (in (A) and (B)), or that one end is projecting out of the right-hand configuration (as in (C)).

Actually, the fitting of one figure to the other does not always involve a one-step transposition, but may require first swinging one end of the left configuration to the orientation of the right, and then following through to the other end of the figure, or even at some stage transposing back to the left-hand figure. (With a little practice, however, it seems that I can, in general, effortlessly and smoothly swing one figure across the page into alignment with the other.) As a result of this interaction between perception and imagination, I know what to say in regard to whether the two figures have the same or a different configuration.

From a phenomenological perspective, then, we have here an interaction between the visual perception of figures, the manipulation of mental representations of these figures in imagination, and verbally encoded thought. If we now examine these capacities further, we will find illustrations of that cluster of features which ground the faculty view of cognition: storage and production facilities with different representational formats, different operating capacities, occupying different roles in relation to behaviour, and responsible for different domains of experience. I will begin with the relations between visual perception and visual imagination. This is a particularly interesting case because we don't get the neat separation of faculties that the initial characterization suggests. Visual perception habitually recruits visual imagination, I will claim, because the two capacities have the same representational format and can communicate extensively. If visual perception and visual imagination are to be counted as different faculties, on other grounds, then they are to be grouped together at some level of analysis on this ground. Their contributions to Gardner's spatial intelligence looks like an obvious candidate here. Even so, I am going to claim that we can properly construe visual perception and visual imagination as distinct faculties, because visual imagination can internally generate transformations on its own representations that are essentially stimulus bound in the case of visual perception, leading to significant cognitive-experiential and behavioural consequences. Next, I want to claim that, our capacity to read conventional notations notwithstanding, the representational format and transformational capacities of these two systems are quite distinct from those of verbally encoded thought, and that communications between these systems and those responsible for generating a subject's speech dispositions are correspondingly limited. The claim that different brain-based systems operate on quite different representational formats provides a central motivation for the faculty view. Here the example from Shepard and Metzler is deliberately chosen, because it locks into the debate about 'descriptional' versus 'pictorial' representational formats for mental imagery. I want to insist on the corollary of limited communication too, however. Finally, I claim that the phenomenology with which I began can be identified with the results of operations carried out in each of these systems, and that it is at least largely via these results that the systems achieve their cooperative efforts, so that these conscious states provide the connecting elements in the conjoint production under analysis. The picture which emerges, if we can generalize from the discussion, is that of a faculty-based cognitive system in which consciousness plays a central role.

2. To the extent that visual imagination may plausibly be regarded as having evolved from visual perception – as being an adaptation of perceptual capacity conferring selective advantage – then we have reason to suppose that the two have extensive and intimate connections. We have reason to suppose, moreover, that the kinds of information bearing structures in which visual imagination encodes or represents spatial configurations – its representational format, in short – are akin to those of visual perception. Plausible evolutionary considerations accord with experimental findings on this point. This is illustrated, for instance, by the Perky effect.[12] Perky found that imaging subjects were unable to report faintly projected stimuli which appeared to have influenced the reports of their mental images, suggesting that the subjects confused the external stimuli with their images. Subsequent investigation has revealed that although failure to report the stimuli is most pronounced when subjects are instructed to form images without prior knowledge that they are being exposed to projected stimuli, as in the original Perky condition, it survives the subject's being informed that stimuli will be present on some trials.[13] Investigation has also shown that although failure to report the stimuli may be in part a result of distraction produced by the activity of imaging itself, this is not the sole factor. In an intermodal study using auditory and visual stimuli, Segal and Fusella[14] found that "the same visual images block visual signals more than auditory signals, and the same visual signals are less detectable when *S* is imagining pictures than sounds". This modal-specific effect suggests that "imagery and perception are similar processes, and visual imagery depends on specific activity in the optic system, while auditory imagery depends on activity in the auditory pathways", and that in the experi-

ments, "imagery functions as an internal signal which is confused with the external signal".[15] In short, the experiments suggest that imagery has strong modal-specific connections and that confusion between image and percept occurs, in part, because the representational format of imagery is akin to that of the perceptual modality to which it is connected.[16]

As Ned Block has pointed out,[17] one can allow this without thereby admitting that the format of either is different from that of, say, verbally encoded thought – that, to use Block's terms, it is 'pictorial' rather than 'descriptional'. I take Block's point. And since the faculty hypothesis which I am proposing requires different representational formats, further argument is required. I will return to this issue shortly. Before doing so, however, I wish to continue with the claim that visual imagination is a separate faculty from visual perception. The argument for this is that, because visual imagination generates operations over representations which in visual perception are generated by stimulus cues, it can operate independently of vision, which, together with the recruitment facilitated by their intimate connections, offers considerable behavioural advantages.

In order to see the selective advantage of visual imagination as an auxiliary to visual perception, let us begin with the phenomenon of zero time delay in visual tracking tasks. As Richard Gregory has pointed out,[18] since there is an inevitable time delay between the onset of a stimulus and any response which it might produce, the fact that in many tracking tasks subjects typically accommodate to a change in the stimulus with zero time delay shows that these accommodations are not produced by the change in stimulus. The simultaneity of stimulus onset and behavioural response is explained, on Gregory's account, by viewing visual perception as a process of selection among competing perceptual hypotheses which, in turn, have sufficient predictive power to adequately control behaviour in advance of the changing stimulus. Regardless of what one makes of Gregory's 'hypothesis' hypothesis, it is obvious that a perceptual system with some measure of predictive power provides a response potential superior to one which merely reacts to stimuli as they present themselves. And the response potential afforded by this predictive power confers selective advantage. It is equally obvious that coupling a disciplined visual imagination to this system produces a considerable advance in predictive power. The capacity that allows us to mentally rotate one member of a Shepard and Metzler stimulus pair onto the other is, more generally, a capacity to visualize how objects will be related when they transport themselves or are transported through the environment. It is a capacity, then, of considerable utility to both the hunter and the builder of shelters. It is also a capacity for envisaging the transformation of natural materials in the fashioning of tools and the manufacture of artefacts. In short, the advent of visual imagination must have opened up a whole new world of spatio-manipulative possibilities of very considerable selective advantage.

My suggestion is that visual imagination is a natural adjunct to visual perception in the development of spatio-manipulative capacities; that its representational format owes much to vision; but that it extends the predictive power of these forms of representation by permitting a massive increase in internal manipulation as against environmental stimulus control. More precisely, it is a capacity which permits both the representation and the transformation over representations of stimulus variables quite independently of the occurrence of stimuli and of transformations over actual stimulus variables in the environment. This at once provides the basis of its selective advantage and the reason why visual imagination, although intimately connected with visual perception, is a capacity essentially separable from it. Its experiential and behavioural consequences provide a proper ground, I would say, for regarding it as a separate faculty.

3. This brings me to the second matter, arising out of relations between visual perception and visual imagination, on the one hand, and verbally encoded thought, on the other. As I said, there are two closely related points at issue here: first, that the representational format and transformational capacities of the former systems are quite distinct from those of verbally encoded thought and, secondly, that communication between these systems is correspondingly limited. Now I cannot hope to prove these points here. In recent years, the first has been subject to considerable debate, largely centring on the interpretation of a body of experimental results obtained by Shepard, Kosslyn, and others, to which I refer the reader.[19] My more modest aim in regard to the first point will be to deflect some well-known attacks on the functional role and experiential reality of mental imagery. As to the second point, the limitation that I see arises out of what would otherwise be a need for a *lingua franca* for the translation of results. Here I shall merely suggest one means of obviating the necessity for such translation.

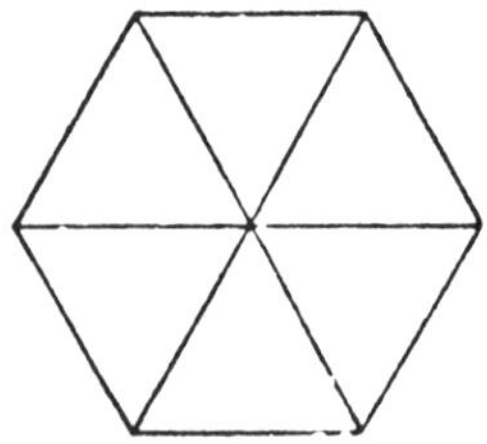

Fig. 2

In arguing against the idea of iconic thought as expressed in the writings of Jerome Bruner, Jerry Fodor[20] has claimed that images cannot by themselves provide a medium of thought, since they are not sufficiently abstract for conveying truth and reference. In order to have the capacity for expressing truth – that *X* is *y*, as it were – an image would need to be able to assign a property to an object. Now an image cannot do that. "Any picture of a thing", Fodor tells us, "will, of necessity, display that thing as having indefinitely many properties; hence pictures correspond (or fail to correspond) in indefinitely many ways to the things that they resemble".[21] And images don't provide the means of selecting which property to assign. So thinking that something is the case cannot be merely entertaining an image. Again, an image could be a thought of something, and hence convey reference, "only when the image is taken to satisfy certain descriptions". Entertaining an image of Figure 2, Fodor tells us, could be thinking of "a pinwheel sort of thing" or thinking of "a cube viewed from one of its corners", but in either case the thought is of its referent only when the image is construed under the appropriate description.[22]

Contrary to Fodor, I am inclined to doubt that semantic features such as truth and reference depend in any general way upon discursive forms of representation. In Fodor's example, the difference between thinking of a cube and thinking of a pinwheel sort of thing could just as easily depend upon a classification defined over images. Thus, Figure 3 suggests alternative disambiguations of Figure 2 by presenting transformations of Figure 2 invariant to different stimulus objects under rotation. The idea is that through being classified with the likes of Figure 3(A), Figure 2 is construed as representing a cube, in that it is classified under an image class, successive members of which would be generated by transformations invariant to presentations of a stimulus cube. Similarly

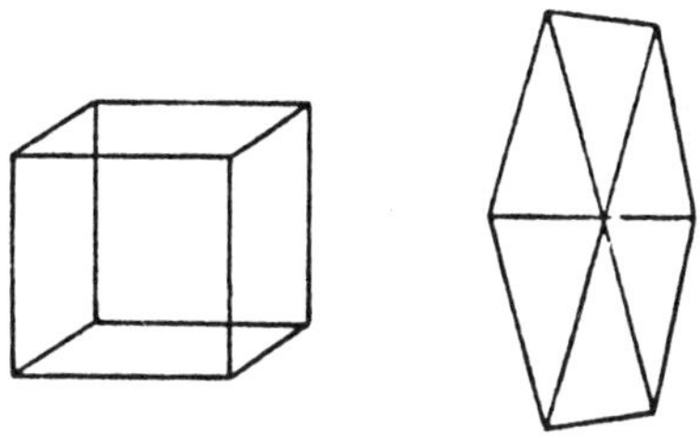

Fig. 3

for the construal of Figure 2 as a pinwheel sort of thing under Figure 3(B). Now if we extend this suggestion to Figure 1, we find that the successive members of the referential image class are generated by transformations that define the rotation. Establishing the identity of a stimulus pair, in that case, is a matter of discovering, by successive transformation, that the comparison figure is represented in the same image class. In other words, coming to think that, under rotation, the left-hand figure is the same as the right-hand figure in, say, Figure 1(A) could be a matter of operating over mental representations of the two figures which themselves turn out to be members of an image class that specifies the one three-dimensional stimulus configuration. So I don't agree that reference and truth, as embodied in 'thinking of' and 'thinking that', are entirely beyond the capacity of imagistic representation. I don't see, in short, why one mode of thought might not be iconic.

On a different tack, Daniel Dennett has attempted to undermine the credibility of subjective judgement in relation to mental imagery.[23] According to Dennett, the subjective judgement made by Shepard's subjects – that they were rotating mental images – involves an erroneous inference reminiscent of the phi phenomenon. That is to say, the 'rotation' is an illusion produced by a rapid succession of discrete changes of image 'orientation'. Moreover, each of these discrete steps really has the character of a judgement, or propositional episode, rather than being truly imagistic or depictional. Dennett says:

(. . .) isn't it the case that if you attend to your experience more closely when you say you rotate the image you find it moves in discrete jumps – it flicks through a series of orientations. (. . .) But now 'look' again. Isn't it really just that these discrete steps are discrete propositional episodes? Now it looks like *this*, but if I imagined it turned *that* much, it would look like *that* . . . ah yes, it would eventually look just like the other one.[24]

Now this is rather interesting. If we suppose for the moment that Dennett's propositional episodes do occur, then they appear to be a commentary woven around a set of intermediate results, all tagged with demonstratives, but remaining themselves propositionally quite unreconstructed. And if the demonstratives tag *images*, which seems to be the obvious suggestion, then this looks like an episode of faculty interaction. It even suggests the possibility that faculties with different representational formats would not have to communicate their results through extensive translation, because much of this would be obviated by the device of demonstrative tags. Actually, I think that this sort of arrangement is widely deployed, and it is the sort of thing that I have in mind when I say that communication between faculties may be rather limited.

Dennett has his response, however, for he elsewhere argues that these demonstratives tag judgments rather than images:

> These judgments have to be about something, and in particular, when I express them, as I did, in such words as 'now it looks like *this*, now it looks like *that*', what can I be ostending or referring to with these demonstratives, if not to some image? But here we are on familiar ground. What good, Wittgenstein reminds us, would ostending an image do here? Consider the duck-rabbit. Now it looks like *this*, now it looks like *that*. What image can you be pointing to here, since the image has not changed at all, though your perception of it has? Well, to an image-that-wears-its-interpretation-on-its-sleeve. But that is no image at all, but an interpretation, a propositional sort of thing, a judgment.[25]

Dennett is certainly right to draw attention to the fact that the mental image is an image-under-interpretation. But he is not so obviously right in insisting that an image-under-interpretation is a 'propositional sort of thing'. As we have seen, an image-under-interpretation could be an image in the context of one or another image class. That is to say, when we imagine that the stimulus figure under rotation would look like *this*, the referent of 'this' might be an image so construed, and not a further propositional element. Moreover, let us suppose that it were a further propositional element. Since it is conscious, it would be a conscious propositional element. If so, it is rather odd that, unlike the accompanying narrative, we experience difficulty in giving it linguistic expression. Even setting these points aside, notice that on Dennett's analysis the case exhibits a more basic feature to be expected on the faculty theory. When we enter upon Dennett's propositional episodes, and imagine that the rotated figure would look like *this*, and now like *that*, the referents of 'this' and 'that' are within consciousness, as I say, and yet external to these propositional episodes. In other words, Dennett's string of propositional episodes and its accompaniment are contemporaneous counterparts evincing a duality of consciousness, just as faculty interaction requires.

In sum, the discussion so far suggests that the representational format and transformational capacities of imagery are quite different from those of verbally encoded thought, that these systems present a duality of consciousness, and that the coordination of effort doesn't require wholesale translation of results between faculties. By way of conclusion to this section, I shall now press the point that we can equate phenomenology with results. For the sake of relief, let's have a change of scenery. If someone asks me to recall my mother's maiden name, for instance, my comprehension of this instruction can very plausibly be regarded as the internal initiator of the sequence which follows. The relevant sequence begins with a comprehended instruction and ends in the mental apprehension of the name. Somehow or other the sequence brings the relevant item to consciousness. Since the item is a person's name, it seems reasonable to suppose that at least a part of the sequence occurs in some area of the brain associated with natural language capacities, and that the result of this processing specifies the content of my consciousness. On this point, failure is at least as instructive as success. The following comes from William James:

> Suppose we try to recall a forgotten name. The state of our consciousness is peculiar. There is a gap therein; but no mere gap. It is a gap that is intensely active. A sort of a wraith of the name is in it, beckoning us in a given sort of direction, making us at moments tingle with the sense of our closeness, and then letting us sink back without the longed for term. If wrong names are proposed to us, this singularly definite gap acts immediately so as to negate them. They do not fit into its mould.[26]

Here processing is in some way blocked so that the desired result is not delivered to consciousness. Yet, as James sets out to argue, we do not draw a complete introspective blank. Something does arise out of the stream of activity. James classifies it as a 'feeling of tendency'. Now this feeling of tendency is often expressed in repeated attempts to say the name or, as James notes, through the rejection of proposed alternatives. We may have some idea of, say, the initial letter or of one syllable and attempt to 'crank-start' the machinery by its repetition. In other cases we test possible candidates against our tacit knowledge for goodness of ness of fit. In either case, I suggest that James' 'gap' in tip-of-the-tongue retrieval is plausibly viewed as a failure of the process to fully furnish the required mental result. 'Crank-starting' seems to be an attempt to produce the result by repeatedly reactivating the procedures through the application of an available partial result. The other strategy appears to involve attempts to fit candidate results to the unrecovered name, by testing them on the procedures which our tacit knowledge suggests should yield the unrecovered name.

The point I wish to stress is that the processes underlying the production of these results and partial results are essentially inaccessible to consciousness. The contents of consciousness are the results so produced. Once brought to consciousness, of course, the name is not only available for verbal report, but may also enter into further cognitions involving further results in a variety of faculties. And these results may not be readily available for verbal report, as we saw in the case of mental imagery. Nevertheless, as that case shows, they may conjointly provide a cognitive basis for behavioural repertoires not available without them.

3. ON THE INFERENCE TO UNCONSCIOUS COGNITION

1. On the present view, consciousness is a feature of behavioural control. The cognitive coordination of a person's behaviour, I have been claiming, is accomplished by a set of faculties, where the results generated by each faculty are identified with consciousness within its representational and operational domain. And given that the coordination of effort here is significantly dependent upon integrating these results, it is dependent upon consciousness. Against this, however, there is a considerable body of experimental research taken to show that subjects are often unaware of the reasoning behind their responses, unaware of shifts in underlying attitudes or values, or of the identity of effective stimuli.[27] Here is a typical example. Gothals and Reckman[28] surveyed high school students on 30 social issues including their attitudes to the use of busing to achieve racially integrated schools, and later divided the students into discussion groups on the basis of their response to this issue. A well-prepared confederate was placed in each group with instructions to argue against the opinions of the group. The confederates proved so successful that most of the pro-busing group actually became anti-busing, while the anti-busing group also moved markedly

in the direction of pro-busing. Yet, when the students were subsequently questioned about their earlier views, those who had been anti-busing recalled their opinions to have been significantly more pro-busing than originally recorded, while most of those who had been pro-busing recalled their view as having been anti-busing from the beginning. No subject reported that they had changed their opinion.

Nisbett and Wilson have assembled an impressive array of results on this issue, and there can be little doubt that subjective reports are fallible in view of other behaviour, and in many cases are little more than unwitting confabulation. Nisbett and Wilson, however, take a stronger line. Having reviewed the literature, they conclude that the introspection of mental processes is a myth. According to Nisbett and Wilson:

> Subjective reports about higher mental processes are sometimes correct, but even the instances of correct report are not due to introspective awareness. Instead, they are due to the incidentally correct employment of a priori causal theories.[29]

These theories, they go on to suggest, may be culturally given, or drawn from the individual's stock of observed stimulus-response correlations or connative associations. In any case, it turns out that a subject's access is not more direct than the experimenter's, and not in general soundly based. In the case of retrospective judgment, for instance, they claim that:

> When people are asked to report how a particular stimulus influenced a particular response, they do so not by consulting a memory of the mediating process, but by applying or generating causal theories about the effects of that type of stimulus on that type of response. These plausibility judgments exist prior to, or at least independently of, any actual contact with the particular stimulus embedded in a particular complex stimulus configuration.[30]

At first glance, this suggests that we are all in fact producing something equivalent to the startling results obtained from the split-brain studies. In the split-brain patients, the left-brain systems responsible for verbal report have no direct access to right-brain perceptual function, so that when they respond to a left-field visual stimulus by selecting the appropriate object with the left hand, they tend to confabulate in order to explain their action. One of Gazzaniga's subjects, for instance, was presented with a left-field visual stimulus word PIPE and a right-field stimulus word CAR, and when asked to select an object from the tray corresponding to the stimulus seen, the left hand selected the pipe while the right hand picked up the toy car.[31] The subject had a ready explanation for this. He saw the word CAR and cars have exhaust pipes, so that he not unnaturally also selected the pipe. The patient denied seeing any stimulus other than CAR. Gazzaniga notes that confabulation in these patients is compulsive. Even when the patients verbally accept the experimenter's account of these behavioural anomalies at the time, they soon return to their rationalizations. The picture is gripping, but it does overstate Nisbett and Wilson's case. For even though they deny access to process, they do not rule out introspection entirely:

> We do have direct access to a great storehouse of private knowledge. (. . .) The individual knows a host of personal historical facts; he knows the focus of his attention at any given point in time; he knows what his current sensations are and what almost all psychologists and philosophers would assert to be 'knowledge' at least quantitatively superior to that of observers concerning his emotions, evaluations, and plans.[32]

The point is that a subject's life-memories, current sensations, perceptual and emotional states, settled aims and values, are resultant psychological *states*, rather than psychological processes. So what Nisbett and Wilson are claiming is that subjects do have direct access to psychological states, but they never have this access to psychological processes.

For the sake of argument, let us suppose that this division of states and processes can be maintained, and also set aside the difficulties involved in relying almost exclusively upon retrospective recall as an index of subjective awareness.[33] To what extent is Nisbett and Wilson's thesis damaging to the view defended here? I shall claim that it is not damaging at all.

2. Let me begin with a couple of preliminary remarks. First, on the faculty view, cognitive consciousness is tied to the results of operations in the faculties, and there is no reason, in principle, why these results should encode or represent the processes from which they are derived. After all, the underlying processes are the *means* by which these results are produced, not the *message*. Secondly, there is no difficulty in holding that states of conscious awareness are intimately involved in cognitive control, and also that subjects sometimes lack the capacity to verbally reconstruct their input. Indeed this is to be expected on the faculty

theory. On this theory cognitive control is a joint effort and requires no omniscient overseer. So there is no reason to expect that the conjoint operations of the faculties are represented anywhere.

As these preliminary remarks indicate, far from there being an inconsistency here, the faculty account looks as if it may well give us an explanation of Nisbett and Wilson's findings. Yet, while I believe that this is so, it is evident that Nisbett and Wilson are themselves operating on a different interpretation of cognitive control. An example will make this clear. Nisbett and Schachter[34] induced subjects to receive steadily increased electric shocks four times the amperage of control subjects by administering a placebo pill, which the subjects were given to suppose would induce the symptoms actually produced by the shock. Now this tends to confirm the experimental hypothesis that the placebo subjects would attribute their symptoms to the pill, and so take more shock. Yet, when questioned, these subjects denied being conscious of making any such attribution:

> Finally, the experimenter described the hypothesis of the study in detail, including the postulated process of attribution of symptoms to the pill. He concluded by asking the subject if he might have any thoughts like those described. Subjects typically said that the hypothesis was very interesting and that many people probably would go through the process that the experimenter described, but so far as they could tell, they themselves had not.[35]

Nisbett and Wilson's conclusion, of course, is that the subjects lacked conscious awareness of their thought processes. Between the data and the conclusion falls an argument, however, and it goes like this. Since the subjects acted in accord with the hypothesis, they must have attributed their symptoms to the pill. But since the subjects did not think this out consciously, they must have done so unconsciously. This seductive argument gains much of its force, I suggest, just because it saves what seems the obvious hypothesis that the subjects took more shock because they attributed their symptoms to the pill. Yet it does this only by the *post hoc* supposition that the subject's behaviour is to be explained in terms of unconscious thought processes of higher cognition. Now, to my way of thinking, this form of explanation is both too intellectualist and systematically incomplete in detail, so that it fails to give proper emphasis to the role of conscious awareness in the production of behaviour, as required on the faculty theory. In order to show this, however, I need to say something about the complex character of those experiences which form around bodily sensation, so that I am about to enter on a brief excursus.

A cooperative effort between distinct capacities can often be revealed even in the seemingly most simple of experiences, and this is so for pain. One might be inclined to think that bodily feelings such as pain are purely sensory affairs. They are sensations, after all. On this view, they standardly involve a sequence of afferent activity initiated at some bodily location and routed through the peripheral nervous system to the brain, where the sequence has its sensory registration. Does the experience of pain involve nothing more than this sort of sensory registration? The answer is that this simple story doesn't stand scrutiny, not even in terms of a plausible taxonomy of the everyday language of pain. This is clearly demonstrated by the work of Melzack and Torgerson.[36] From earlier taxanomic studies and a general examination of the clinical literature, Melzack and Torgerson produced an alphabetical list of 102 words describing painful qualities, and in subsequently administered tests were able to show that subjects expressed very substantial agreement in categorizing these terms. In brief, they arrived at three major categories for describing pain. First, there are those terms that can be classified as *sensory-discriminative*, as for example when we speak of the pain as stabbing, sharp, burning, shooting, throbbing, or stinging. Then there are *cognitive-evaluative* terms. For instance, we say that a pain is mild, strong, or unbearable. Finally, there are those *motivational-affective* words, as when we say that a pain is sickening, tiring, fearful, terrifying, and so on. Agreement also extended to placing words in subdivisions in two of these classes, so that sensory-discriminative qualities were taken to be either temporal, spatial, thermal, in some sense bright or dull, or as having to do with pressure of one sort or another, while motivational-affective qualities were agreed to belong to the categories of tension, fear, punishment, or autonomic response. Finally, it was found that most words were consistently evaluated on a five point intensity scale from Mild to Excruciating.

Clearly, Melzack and Torgerson's taxonomy demonstrates that, given the ways in which we are inclined to describe and categorize pain, it cannot be that our experience shows pain to be a simple sensory phenomenon. On the contrary, the ways in which we describe our experience suggest that it is a composite phenomenon, incorporating sensory, evaluative, and affective dimensions.

There is also a good deal of clinical, experimental, and sociological evidence for the responsiveness of affective and sensory dimensions of the pain experience to the subject's cognitive evaluations.[37] The administration of placebos, for instance, appears to be effective in reducing severe pain in about one third of post-operative patients, suggesting that post-operative pain can have a significant evaluative component tied to the subject's beliefs. A frequently cited case is Beecher's comparative study of pain occasioned by wounds in battle and pain resulting from surgery. Only around one third of severely injured soldiers in combat hospitals required morphine, he notes, by comparison with about four-fifths of their civilian counterparts. Beecher suggests that this is largely explained by differences in the perceived significance of their respective circumstances. To arrive at a combat hospital meant relief and safety for the badly wounded soldier, whereas the civilian's major surgery was viewed with deep distress. The sensory and affective elements are seen to be distinct, in addition, from cases where affect is diminished or effectively eliminated while sensory discrimination remains intact. For instance, it is often commented that patients who have undergone frontal lobotomy for the alleviation of chronic severe pain apparently continue to experience the pain but say that they no longer mind it. The procedure radically alters the patient's affective response, but not it seems the sensory component of the pain. A similar phenomenon seems also to occur in cases of congenital insensitivity to pain. These people can discriminate pricking, temperature, and pressure intensities at noxious levels even though there is no corresponding aversive response. These clinical studies, I suggest, argue for a faculty theory of pain, in which the experience of pain is a composite of the results of processing in a number of systems with different computational resources. End of excursus.

Now, given this, a more plausible interpretation of the behavioural controls operating in the case of Nisbett and Schachter's subjects goes roughly as follows. When the subjects were told that the pill would cause respiritory irregularity, palpitations, tremor, and so on, this created an expectation consonant with the sensory consequences of electric shock. In addition, the subjects were expecting to be exposed to shock. In the event, of course, the subjects would have no ready means of determining to what extent their symptoms were due to the pill or to the shock. It is likewise difficult for us to tell to what extent their expectation or mental set in relation to these symptoms was dissociated from their expectations of shock. Yet, insofar as the subjects were concerned with the effects of the shock rather than those of the pill, any dissociation would have the effect of inducing the subjects to place a lower than usual cognitive evaluation upon the symptoms. Now with this cognitive-evaluative element in place, the subjects entered the experimental phase, in which their attention was directed to the shock. As Nisbett and Wilson report, in response to the question "While you were taking the shock, did you think about the pill at all?" the typical response was "No, I was too worried about the shock". Even so, by prior arrangement, as we saw, the subjects were already disposed not to respond normally in evaluating the sensory experience induced by its onset. As a result of this the subjects experienced a weaker affective reaction, were consequently less highly motivated to end the trials, and so took more shock than the controls.

On this interpretation, the experimental subject's behaviour is not to be explained simply by hypothesizing a chain of unconscious inference. Instead it is claimed that the experimental setup effectively modified the cognitive-evaluative and motivational-affective dimensions of their experience of pain. And it was because their experience of pain was affected in this way that they took more shock. The conviction that unconscious thought processes *must* occur, I suggest, is often based on nothing more than an intellectualist prejudice. The intellectualist, on the view here ascribed, is someone who tends to cast all mental operations in the mould of propositional thought. So it is said that Nisbett and Schachter's subjects must have *reasoned* that their symptoms were attributable to the pill. And when it turns out that the subjects cannot report this piece of reasoning, it is said that they must have done it unconsciously. Yet the intellectualist's initial assumptions about what must have occurred are by no means mandatory, as we have seen. Indeed, *prima facie* they present us with an excess of ratiocination. Once we suppose that these assumptions might misconstrue the mental operations underlying performance, however, then we raise the possibility that the experimenters fail to elicit the expected retrospective reports not because these reasonings were unconscious but because they did not occur. We raise the possibility, moreover, of accounting for the experimental findings in terms that refer us back to the subjects' experience. The latter course, I take it, is to be preferred. It makes the saving hypothesis of unconscious reasoning otiose by explaining the behaviour, both in the experiment and at post-test interview, in terms of experience. This preferable course has been made available, however, only because we have recognized that the subject's experience conforms to a faculty model. While a faculty theory of pain is sustained on other grounds, the fact that it allows us to take this course, and to deny the intellectualist inference, itself provides an argument in favour of a faculty theory.

In these notes toward a faculty theory of cognitive consciousness I have attempted to suggest only the general direction that the theory would take. There has been little reflection upon the internal structure of faculties and their cooperative mechanisms. No list of faculties has been attempted, and the representational diversity of these systems was barely sampled. Moreover, since the faculty theory would be no less than a theory of mind, it requires development on many fronts that I have not in the least mentioned. Some developments are not hard to envisage. As with 'bodily feelings', emotion looks susceptible to a faculty analysis; and attention might fruitfully be seen in terms of temporary cooperative structures formed when faculties recruit one another. The difficult task is to show that no work remains to be done by 'higher cognition' independently of the cooperative efforts of faculties, and it is here that the whole project seems most likely to founder. Setting all of this aside, there is a central question which I have not addressed because it requires a detailed discussion of its own. What precisely is it about the results of operations in the faculties in virtue of which they are conscious? Here are three possible preliminary answers. (1) The operations within a faculty don't come to consciousness because of their rapidity. (2) Consciousness requires the formation or reactivation of a mental representation, and mental representation in this explicit form occurs only in the outputs of the faculties. (3) Consciousness is a relational rather than an intrinsic property, and only the access relations between the faculties provide the basis for it. I will leave it to the reader to judge whether any of these possibilities is worthy of serious consideration.

NOTES

[1] See Jerry Fodor (1983) *The Modularity of Mind*, MIT, and Howard Gardner (1983) *Frames of Mind: The Theory of Multiple Intelligences*, Paladin.

[2] See my paper (1988) 'Modularity, Rationality, and Higher Cognition', *Philosophical Studies* **LIII**, pp. 279–294.

[3] See Gardner, *op. cit.* (note 1)
[4] Michael Gazzaniga and Joseph Le Doux (1978) *The Integrated Mind*, Plenum Press, pp. 13–17.
[5] The literature here is extensive. Useful sources are R. W. Sperry (1974) 'Lateral Specialization in the Surgically Separated Hemisphere, F. O. Schmitt and F. G. Worden (eds.), *The Neurosciences: Third Study Program*, MIT Press, pp. 5–19, and M. S. Gazzaniga (1970) *The Bisected Brain*, Appleton-Century-Crofts.
[6] R.W. Sperry (1984) 'Consciousness, Personal Identity and the Divided Brain', *Neuropsychologia* **XXII**, pp. 661–673. See p. 662.
[7] R.W. Sperry (1968) 'Hemisphere Deconnection and Unity of Conscious Awareness', *American Psychologist* **XXIII**, pp. 723–733, as quoted by Roland Puccetti (1973) in 'Brain Bisection and Personal Identity', *British Journal for the Philosophy of Science* **XXIV**, pp. 339–355, on p. 343.
[8] Puccetti, *op. cit.* (note 7), p. 343.
[9] See *op. cit.* (note 4.) pp. 155–159, for interesting remarks on this in relation to cognitive dissonance. See also Section 3 below.
[10] See *op. cit.* (note 6), p. 666.
[11] The *locus classicus* is R.N. Shepard and J. Metzler (1971) 'Mental Rotation of Three-Dimensional Objects', *Science* **CLXXI**, pp. 701–703. (Figure 1 is reproduced by permission of the authors and *Science*. (Copyright 1971 by the American Association for the Advancement of Science.)
[12] C.W. Perky (1910) 'An Experimental Study of Imagination', *American Journal of Psychology* **XXI**, pp. 422–452.
[13] S.J. Segal and P.E. Gordon (1969) 'The Perky Effect Revisited: Blocking of Visual Signals by Imagery', *Perceptual and Motor Skills* **XXVIII**, pp. 791–797.
[14] S.J. Segal and V. Fusella (1970) 'Influence of Imaged Pictures and Sounds on Detection of Visual and Auditory Signals', *Journal of Experimental Psychology* **LXXXIII**, pp. 458–464.
[15] *Ibid.* pp. 458–459 and 463.
[16] For further results to the same effect see Roger Shepard (1978) 'The Mental Image', *American Psychologist* **XXXIII**, pp. 123–137, and R. A. Finke (1980) 'Levels of Equivalence in Imagery and Perception', *Psychological Review* **LXXXVII**, pp. 113–139.
[17] Ned Block (ed.) (1981) *Imagery*, MIT Press, p. 10.
[18] Richard Gregory, 'Perceptions as Hypotheses', (1974) *Philosophy of Psychology*, S. C. Brown (ed.), Macmillan, pp. 202–203.
[19] See Ned Block's *Imagery* for a useful collection of this material.
[20] Jerry Fodor (1976) *The Language of Thought*, Harvester.
[21] *Ibid.* p. 181.
[22] *Ibid.* p. 182.
[23] D.C. Dennett (1978) *Brainstorms*, Bradford Books, pp. 167–169. For a more complete account along the following lines, see my 'Propositions About Images', forthcoming in *Philosophy and Phenomenological Research*
[24] *Ibid.* p. 168.
[25] D. C. Dennett, 'On the Absence of Phenomenology' (1979) *Body, Mind and Method*, D. F. Gustafson and B. L. Tapscott (eds.), Reidel, p. 105.
[26] Quoted in *op. cit.* (note 23), p. 160.
[27] An influential surevey of this material is Richard E. Nisbett and Timothy DeCamp Wilson (1977) 'Telling More Than We Can Know: Verbal Reports on Mental Processes', *Psychological Review* **LXXXIV**, pp. 231–259.

[28] G.R. Goethals and R.F. Reckman (1973) 'The Perception of Consistency in Attitudes', *Journal of Experimental Social Psychology*, **IX** pp. 491–501.
[29] See *op. cit.* (note 27), p. 233.
[30] Ibid. p. 248.
[31] M.S. Gazzaniga and C.S. Smylie, 'What Does Language Do for a Right Hemisphere', (1984) *Handbook of Cognitive Neuroscience*, M. Gazzaniga (ed.) Plenum Press, pp. 199–209.
[32] See *op. cit.* (note 27), p. 255.
[33] On these and other difficulties see K. Anders Ericsson and Herbert Simon (1980) 'Verbal Reports as Data', *Psychological Review* **LXXXVII**, pp. 215–251, and Eliot R. Smith and Frederick D. Miller (1978) 'Limits on Perception of Cognitive Process: A Reply to Nisbett and Wilson', *Psychological Review* **LXXXV**, pp. 355–362.
[34] R.E. Nisbett and S. Schachter (1966) 'Cognitive Manipulation of Pain', *Journal of Experimental Social Psychology* **II**, pp. 227–236.
[35] See *op. cit.* (note 27), p. 237.
[36] R. Melzack and W.S. Torgerson (1971) 'On the Language of Pain', *Anesthesiology* **XXXIV**, pp. 50–59.
[37] For references and discussion of the following and other cases see Ronald Melzack (1973) *The Puzzle of Pain*, Basic Books. Suggestions about plausible physiological substrates for what follows are canvassed in Chapter 6.

School of Philosophy
University of New South Wales, Sydney, Australia

MICHAEL A. ARBIB

MODULARITY, SCHEMAS AND NEURONS: A CRITIQUE OF FODOR*

1. MODULES IN THE BRAIN

It is a standard notion that a complex system may be analyzed by being decomposed into a set of interacting subsystems. Such a decomposition succeeds insofar as we can understand the relation between the inputs and outputs of each individual subsystem, and insofar as the interactions between the subsystems can be explained via suitable connections between various of their inputs and outputs, without further analysis of variables internal to the subsystems. Suci a decomposition is *structural* to the extent that the subsystems can be mapped onto physical substructures of a physical structure embodying the overall system. In this section, I show that neuroscientists have long sought structural decompositions of the brain, and in some cases referred to the physical substructures as *modules*. Recently, Fodor has popularized the use of the term 'module' to denote a unit in a functional decomposition of a cognitive system, but a subsystem that meets constraints beyond those specified above. I shall argue that Fodor's analysis of cognitive systems is flawed and that the restrictions he introduces are not useful. Consequently, I shall use the term 'module' as a synonym for the term 'subsystem' defined above.

The work of the nineteenth century neurologists led us to think of the brain in terms of large interacting regions each with a more or less specified function, and this localization was reinforced by the work of the anatomists at the turn of the century who were able to subdivide the cerebral cortex on the basis of cell characteristics, cytoarchitectonics. It was at this same time that the discoveries of the neuroanatomist Ramon y Cajal and the neurophysiologist Sherrington helped establish the neuron doctrine, leading us to view the function of the brain in terms of the interaction of discrete units, the neurons. The issue for the brain theorist, then, is to map complex functions, behaviors, patterns of thought, upon the interactions of these rather large entities, anatomically defined brain regions, or these very small and numerous components, the neurons. This has led many neuroscientists to look for structures intermediate in size and complexity between brain regions

Peter Slezak and W.R. Albury (eds.), Computers, Brains and Minds, 193–219

and neurons to provide stepping stones in an analysis of how neural structures subserve various functions. One early example was the Scheibels'[1] suggestion that the reticular formation could be approximated by a stack of 'poker chips' each incorposating a large number of neurons receiving roughly the same input and providing roughly the same output to their environments. This modular decomposition of the reticular formation provided the basis for the Kilmer and McCulloch model, RETIC.[2]

In another direction, the theoretical ideas of Pitts and McCulloch[3] combined with the empirical observations of Lettvin, Maturana, McCulloch and Pitts[4] on the frog visual system to suggest that one might think of important portions of the brain in terms of interacting layers of neurons, with each layer being retinotopic in that the position of neurons in the layer was correlated with position on the retina, and thus in the visual field. Powell and Mountcastle[5] working in somatosensory cortex, followed by Hubel and Wiesel[6] working in visual cortex, established the notion of the column as a 'vertical' aggregate of cells in cerebral cortex, again working on a common set of inputs to provide a well-defined set of outputs. With all these considerations, the notion of the brain as an interconnected set of *modules* – intermediate in complexity between neurons and brain regions – was well established. For example, Szentágothai and Arbib,[7] in their monograph on *Conceptual Models of Neural Organization*, wrote that:

> The concept of a modular structure or arrangement of the neuropil has two basic sources:
> (1) a more indirect one from the notion that neuronal networks ought to be subdivided into distinct functional units; and
> (2) a direct one from observation of the neuropil (. . .). Recent anatomical data on the cerebral cortex suggest the existence of both a fine grain and coarse grain of modular organization.

Subsequently, Mountcastle,[8] in an essay titled 'An Organizing Principle for Cerebral Function: The Unit Module and the Distributed System', wrote that:

> The large entities of the brain we know as areas (or nuclei) (. . .) are themselves composed of replicated (. . .) modules (. . .). Each module processes information from its input to its output and in that processing imposes transforms determined by the general properties of the entity and its extrinsic connections. (. . .) Closely linked (. . .) subsets of modules (. . .) form precisely connected but distributed systems. (. . .) A single module of an entity may be a member of several (but not many) such systems.

With this, it is clear that the concept of a 'module' is well established within neuroscience as a structural entity, but our task in this paper is to confront it with the notion of a module as a functional entity, as developed, for example, in the elegant monograph *The Modularity of Mind* by Jerry Fodor. To see this, it will help to distinguish what has been called *top-down* brain theory from *bottom-up* theory, and see how they are brought together in what might be called 'middle-out' brain theory. Top-down theory is essentially functional in nature, in that it starts with the isolation of some overall function, such as some pattern of behavior or linguistic performance or type of perception, and seeks to explain it by decomposing it into the interaction of a number of subsystems. What makes this exercise brain theory as distinct from cognitive psychology is that the choice of subsystems is biased in part by what we know about the function of different parts of the brain, as obtained for example by analysis of the effects of brain lesions, so that there is some attempt to map the subsystems onto anatomical regions.

In bottom-up brain theory, the emphasis tends to be on neural circuits. Given a specific set of neurons, the attempt is to use them to implement a given function, or to analyze given circuitry to determine what functions it can perform. Clearly, the primary sources of data for such bottom-up brain theory come from both neuroanatomy – in looking at the detailed interconnections of neurons – and from neurophysiology – in studying the behavior of the network under varying conditions.

In its full development, brain theory incorporates a cycle of both top-down and bottom-up modelling. An attempt is made to map functional units onto brain regions, constrained for example by lesion studies. But this map is not one-to-one, and a further constraint is to try to implement regional functions via neural networks meeting anatomical and physiological constraints. New information may not only yield different suggestions as to how circuitry may subserve a given function, but – as we shall see well exemplified in section 4 – may lead to changes in our ideas about how functions are distributed around the regions of the brain. In particular, we shall see that there is a continual tension between functional decomposition and structural decomposition, and that in general a given 'functional module' may be subserved by the interaction of several 'brain modules', and that a given 'brain module' may be involved in subserving a number of different functions – as was indeed suggested by the quote from Mountcastle above.

2. FODOR'S TAXONOMY OF THE MIND

For Fodor, a computational theory of cognitive science seeks a set of mechanisms each of which provides a characteristic pattern of transformations of mental representations.[9] For him, a computational process is by definition syntactic.[10] Motivated by Gall's view of faculties, Fodor seeks cognitive mechanisms which form *modules* in the sense that they are[11]

(i) Domain-specific,
(ii) Innately specified,
(iii) Associated with distinct neural structures, and
(iv) Computationally autonomous.

In the present paper, I shall not address claims (ii) and (iii), but shall restrict myself to setting forth my argument that cognitive science will not be served by an emphasis on modules which are domain-specific and computationally autonomous.

The term 'domain-specific' seems to refer to gross modalities like 'vision' or 'language' rather than Gall's faculties or the 'domains' or 'microworlds' of current AI research. The key concept is that of *computational autonomy*: modules "do not share, and hence do not compete for, such horizontal resources as memory, attention, judgement, intelligence, etc.".[12] More generally, the modules considered by Fodor are *informationally encapsulated*,[13] which means that, while there may be internal feedback between the representations within the module, these internal representations are not involved in paths to or from external modules. Moreover, Fodor argues that there is no external feedback path whereby the output of the module can affect its input (Figure 1).

Fodor argues for an exclusive but not exhaustive functional taxonomy of the mind as divided into transducers, input systems, and central processors.[14] He views the input systems as delivering representations that are most naturally interpreted as characterizing the arrangement of things in the world,[15] and then lumps perceptual systems with language as constituting the input systems, with the note[16] that what underwrites the correspondence between, say, visual stimuli and distal layouts are (roughly) the laws of light reflectance, whereas what underlies the correspondence between token utterances and distal layouts is (roughly)

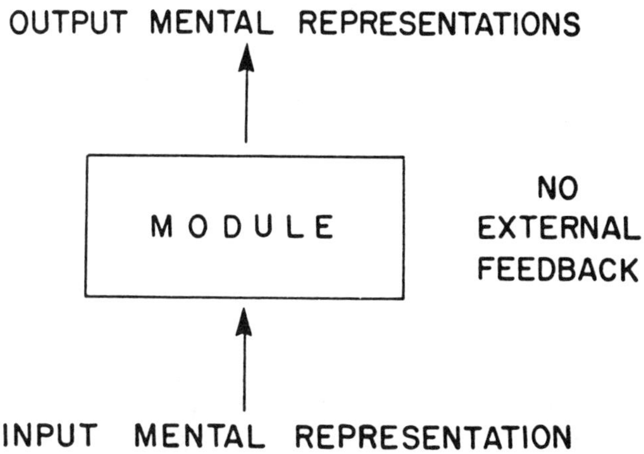

Fig. 1. A key property of a module is that it be computationally autonomous and informationally encapsulated. Moreover, its input cannot depend upon its output via paths internal to the cognitive system in which it is embedded.

a convention of truth-telling which makes it possible to infer from what one hears to the way that the world is.

Fodor's central hypothesis about the mind is that the input systems (perceptual systems and language) are precisely the modules of the mind, "domain-specific computational systems characterized by informational encapsulation, high speed, restricted access, and neural specificity".[17] "The input systems being informationally encapsulated compute representations of the distal layout on the basis of less information about the distant layout than the organism has available. Such representations want correction in the light of background knowledge and of the simultaneous results of analysis in other domains. Call the process of arriving at such corrected representations 'the fixation of perceptual belief'."[18]

Fodor then argues[19] that an interface between perception and utilities must take place *somewhere* if we are to use the information that input systems deliver in order to determine how we ought to act. Fodor then argues that these central processes for thought and problem solving are unencapsulated and so are not plausibly viewed as modular.[20] Although Fodor gives no diagrams in his book, his theory of the mind is, I think fairly, captured in Figure 2.

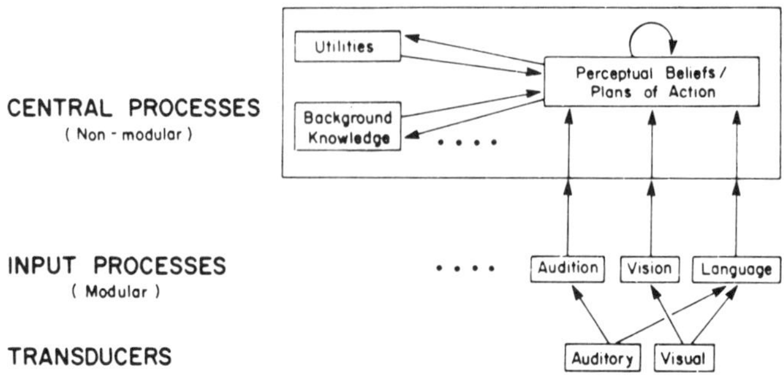

Fig. 2. Fodor's Taxonomy of the Mind. The arrows linking the central processes have no significance other than to indicate the promiscuity of data-flow which renders the central processes non-modular. By contrast, the lack of feedback paths around each input process is crucial to Fodor's theory.

Some comments are in order. First, we reiterate that Fodor's use of 'domain' seems to refer to 'language or a sensory modality'. Second, Fodor gives no analysis of action and motor control, so it is open as to whether or not he would regard these as modular, e.g., with modules for locomotion, manipulation, and speech. But third, and this provides the starting point for our critique, Fodor is silent about which transducers feed the language module. Ignoring braille, I have arbitrarily shown auditory and visual transducers providing this input. But once we insist that, e.g., the visual transducers deliver input to vision, audition and language without cross-talk between these modules (for such cross-talk would break the posited informational encapsulation), the transducers no longer 'merely' transduce sensory signals, but carry out some moderately sophisticated transformations of their own, as sketched in Figure 3. For example, since the visual recognition of curves is helpful both in seeing objects and in recognizing letters, we must either make this curve analysis part of the visual transducer or we must see it as constituting an input module which is not domain-specific. In either case, the distinction between transducers and input processes now seems less comfortable, and will become more so if we try to refine the language, vision, and touch modules in a way that will accomodate the learning of braille and sign language, for now the intermixing of language with touch and motor control makes claims for the informational encapsulation of the modules of Figure 2 well-nigh untenable.

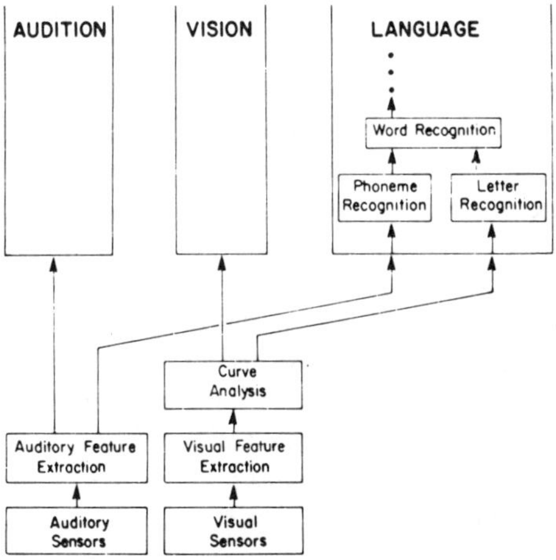

Fig. 3. A hypothetical refinement of the transducers and input processes of Figure 2.

However, my fundamental point is that Fodor's modules are *too large*. It is clear from Figure 3 that a computational theory of cognition must use a far finer grain of analysis than that offered by Fodor (Figure 2). Fodor offers *big* modules, argues vociferously that they are computationally autonomous, and despairs at the problem of explaining the central processes, since they are not informationally encapsulated (more of this in section 6). By contrast, my approach is to analyze the brain in terms of smaller components. Since the interactions between these components play a vital role in my models, the case for autonomy of large modules becomes less plausible – a useful measure of parsimony, rather than a fundamental principle.

In the next three sections I shall try to exemplify this approach to the study of modularity and interaction of brain regions by looking at a number of models of visual perception and of visuomotor coordination. In the first two and one half of these sections, I will be guided by the slogan set forth in *The Metaphorical Brain*[21] that "The brain should be modelled in terms of distributed action-oriented computation in layered somatotopically organized machines". The term 'distributed' emphasizes the notion that the brain is made up of many different systems which are simultaneously active, and so is not to be modelled in terms of serial computation, in which one localized operation is conducted at a time.

Secondly, the notion that the brain is 'action-oriented' emphasizes that we should not think of, for example, vision purely in terms of rendering some sort of objective representation of the visual world within the brain, but rather should ask how that visual representation can provide information that is relevant to the activity of that particular organism. For example, we shall see that the visual system of a frog is quite different from the visual system of a human, even though we shall at the same time strive to find general principles which will help us understand what is involved in the structure of these systems. Finally, the slogan that the brain is 'a layered somatotopically organized machine' will correspond to the claim, made in section 1, that many parts of the brain can be analyzed in terms of the interaction of layers of similar components. However, in the last part of my discussion of vision in section 5, I will point out that there are processes that do not seem to be naturally thought of in terms of somatotopic or retinotopic computation in a layered structure. This will serve to introduce the notion of *schema* as a fine-grained functional module, corresponding for example to the knowledge required to recognize a house or to grasp a mug. In this paper, I shall offer no data on the extent to which schemas may be considered as psychologically or neurophysiologically verifiable, but simply show their computational role in a model of high-level vision. In other writings, I have presented the notion in the context of toad visuomotor coordination,[22] developed its use for explaining skilled manual behavior,[23] and explored with my colleagues the use of schemas in developing various models of language.[24]

3. DEPTH PERCEPTION

A chapter in brain theory usually starts by trying to characterize some overall function which one thinks might constitute a 'brain module' in that one might hope to find specific brain circuitry devoted to its implementation. For example, it seems reasonable to think that 'solving differential equations' would not be a suitable module for such a study, and that our eventual understanding of its brain implementation will come by seeing how cerebral circuitry capable of implementing many different human skills can be tuned by experience and instruction to subserve this particular mathematical ability. By contrast, we have every reason to think of vision as having special circuitry, from the retina of the eye to the tectum of the midbrain and the lateral geniculate nucleus of the thalamus up to a number of regions designated as visual cortex. In fact, the work of the last twenty years has taken us far beyond this conception, and we now talk of the 'many visual systems' finding that there are now tens of anatomically distinct brain regions each subserving distinctive functions of vision. Within this context, then, we may seek to find natural functional modules which subserve part of the task of vision, thus focusing our attention on a more restricted system of neural circuitry. One such natural candidate is *depth perception* – which enables us to determine the world in terms of objects located at various distances from us. This is clearly very important from an action-oriented perspective, since the way in which we will interact with objects will depend crucially on how near or far they may be.

It is a familiar experience from using stereo viewers that the view of a three dimensional scene presented at the left eye differs from that presented to the right eye, and that the disparity or displacement between these two images provides the crucial cue as to the distance in space from which they come. A key concern from the nineteenth century was whether depth perception comes before or after pattern recognition. Is it that the brain takes the image from each eye separately to recognize, for example, a house therein, and then uses the disparity between the two house images to recognize the depth of the house in space; or is it that our visual system matches local stimuli presented to both eyes, thus building up a depth map of surfaces and small objects in space which provides the input for perceptual recognition? It was the great achievement of Bela Julesz to invent the method of random dot stereograms, providing stereo pairs each of which contains only visual noise, but so designed that the visual noise was correlated. Patches of random light and dark presented to one retina were identical to, but at varying disparities from, patches of light and dark presented to the other retina. Julesz found that people were in fact able to carry out the appropriate matching to see surfaces stippled with random patterning at varying depths in space. In other words, without precluding that some depth perception could follow pattern recognition, he did establish that the forming of a depth map of space could precede the recognition of pattern. He offered a model of this process in terms of cooperative computation involving a somewhat Heath Robinson array of magnetic dipoles connected by springs.

For the brain theorist the issue was thus raised: "Could the depth map be computed by a cooperative process involving realistic neurons?", and the data on what constituted realistic neurons was provided by the work of Barlow and various co-workers. One of the first papers to address this issue was that by Arbib, Boylls and Dev[25] who built a neural net cooperative computation model for building the depth map "guided by the plausible hypothesis that our visual world is made up of relatively few connected regions". The neural manifold of this model had cells whose firing level represented a degree of confidence that a point was located at a corresponding position in three dimensional space. The neurons were so connected via inhibitory interneurons as to embody the principle that cells which coded for nearby direction in space and similar depth should excite each other, whereas cells which corresponded to nearby direction in space and dissimilar depth should inhibit each other. It was shown by computer simulation by Dev[26] and later established by mathematical analysis by Amari and Arbib[27] that this system did indeed yield a segregation of the visual input into connected regions. Later, a variant of this model was published by Marr and Poggio,[28] and in subsequent writings Marr[29] took the "plausible hypothesis that our visual world is made up of relatively few connected regions" and showed how it could be developed into an elegant mathematical theorem relating the structure of a depth perception algorithm to the nature of surfaces in the physical world.

However, it must be noted that the fact that a brain mechanism is employed to implement a particular functional module by the brain of one animal does not imply that this is the mechanism used by a different species. It is known that frog and toad can snap with moderate accuracy at prey located in the monocular visual field, and this led Ingle to hypothesize that for the frog it was accomodation, focal length information for the lens, that subserved depth perception. More detailed experiments by Collett, involving placing on the nose of the toad spectacles which used either prisms or lenses, showed that the story was more complex. Collett was able to show that the monocular animal did indeed use accommodation as the depth cue, but that in an animal with prey in the binocular field, the major depth cue was disparity, with accommodation cues exerting perhaps a six percent bias on the depth judgement based on disparity. This led Donald House[30] in my laboratory to suggest that another route to solving the problem of false minima was to use accommodation cues to bias disparity. The key observation here is that the two sources of depth information provide complementary information. House's model uses a variant of the Dev model to refine the depth estimates within each depth field, but adds mutual coupling between the models so that activity localized within one map helps increase activity at the corresponding locus on the other map.

In concluding this section, we make several points. Firstly, that there is no unique algorithm for solving a given problem, in part because many different sources of information can be employed. Secondly, any one source of information will be incomplete, and it will often take the skillful deployment of several sources of information to yield a far better estimate than could be gained by using one source alone. And thirdly, we have seen how these considerations lead us to a far finer modular decomposition than that offered by Fodor (and which we have diagrammed in Figure 3).

4. VISUOMOTOR COORDINATION IN FROG AND TOAD

In this section, we will return to the action-oriented view of visual perception by looking at something of the behavioral repertoire of frog and toad, and asking what type of visual system could make the appropriate information available. In other words, in the title of Lettvin, Maturana, McCulloch and Pitts,[31] we seek to determine "What the Frog's Eye tells the Frog's Brain". The key to the Lettvin et al. paper was to use naturalistic stimuli such as small moving objects similar to the frog's prey of flies and worms. Indeed, extending the results of Barlow[32] they found that there were cells in the retina that seem to serve as 'bug detectors', sending information back retinotopically to the tectum, the visual midbrain. They found that the retina not only sends back a map of where the bugs might be, but also sent back other maps in spatial register, including a map of where there were large dark moving objects in the visual field. It was also known that an animal without a tectum would not engage in prey-acquisition behavior. This led to the first model of prey-acquisition which involved two modules, the retina was the prey recognition module, the tectum was the module subserving motor control for prey-acquisition.

Peter Ewert conducted a number of experiments which led to a somewhat more subtle view of the situation. Firstly, he observed that removing the pretectum, a small brain region just in front of the optic tectum, yielded a toad that would snap even at large moving objects. This immediately leads to a different functional decomposition of the brain into modules. Where before we had a module for recognizing prey, now it would seem that we have two distinct modules – one for recognizing moving objects, and one for recognizing large moving

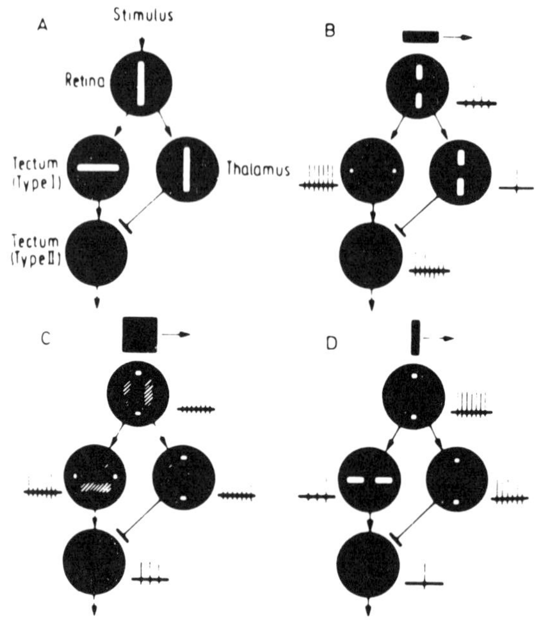

Fig. 4. A lumped model of prey-predator discrimination: a worm filter provides excitatory input to an output cell which also receives inhibitory input from an antiworm cell. The result is that (B) a worm provides a potent response; (C) a square provides an intermediate response; while (D) an antiworm yields little or no response.[34]

objects. It is inhibition of prey-acquisition by the latter that yields the appearance of a module for the detection of small moving objects *per se*. Further work reviewed by Ewert[33] replaced the use of naturalistic stimuli with precise patterns to quantify the animal's response. He found that rectangles elongated in their direction of movement (worms) were stimuli of increasing effectiveness with increasing length; but an anti-worm (a rectangle moving orthogonally to its long axis) quickly became an ineffective stimulus as it was elongated; while a square had an intermediate effect, with bigger squares being more effective stimuli until a certain critical size after which they became less effective stimuli. On the basis of this, Ewert and von Seelen[34] came up with the model of Figure 4 in which the retina was seen as a preprocessor of visual stimuli, the pretectum was a module for anti-worm recognition supplying inhibition to the motor output system, while the tectum was seen as comprising two modules, one for worm recognition and one for motor control for prey-acquisition. This motor control module was excited by the worm recognition module and inhibited by the anti-worm recognition module.

Cervantes, Lara and Arbib[35] carried this work further by going from the modules defined purely by an overall linear response function to modules defined by the detailed interaction of neurons. Lara, Arbib and Cromarty[36] had built on the neuroanatomical observations of Székely and Lázár[37] to define the 'tectal column' as a basic cluster of cells working together at one locus in the tectal layers. Cervantes *et al.* then modeled the tectum as in Figure 5 as an array of such columns,

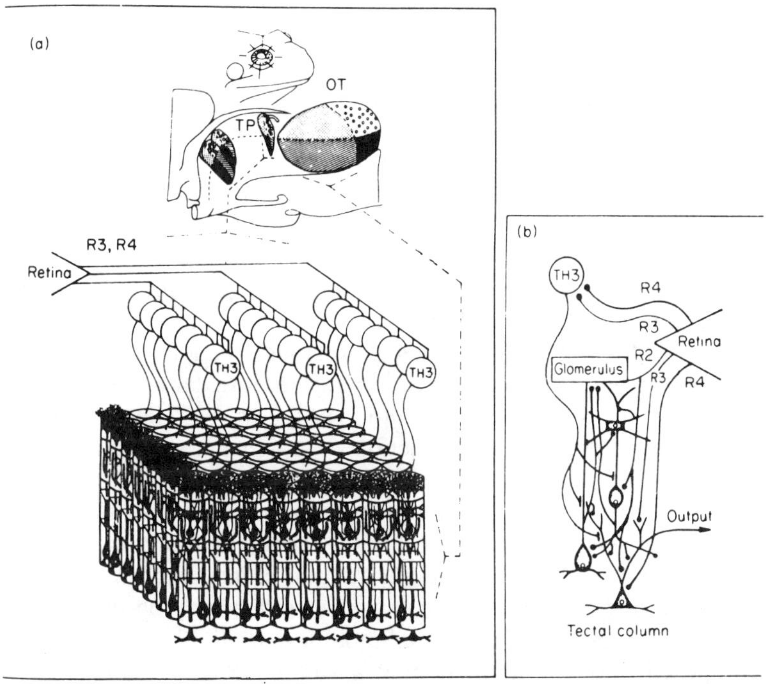

Fig. 5. Interactions among retina, optic tectum and pretectum. The retina sends fibres in a retinotopical fashion to both optic tectum (class R2, R3 and R4), and pretectum (class R3 and R4). (a) TH3 neurons also project retinotopically to the optic tectum. For simplicity we only show the projection of three rows of TH3 cells projecting upon the tectal columns. (b) A closer look at the interactions among retinal, tectal and pretectal cells. The TH3 cell of the pretectal column inhibits LP, SP and PY of the tectal column corresponding to its retinotopic projection.[38]

interfaced with an array of pretectal 'anti-worm detector' neurons, all driven by suitable classes of retinal input, and then showed that this model could explain the spatial and temporal properties of prey-predator discrimination in terms of neural interactions. A fuller analysis of these and other models, together with commentaries by a number of peer reviewers, appears in Arbib.[39]

Such studies not only establish that cognitive science needs a modular decomposition finer than that of Fodor's, but also establish that the choice of modules can be constrained by the data of neuroscience. Since my examples are taken from visuomotor coordination they are specific to the domain of vision, but they do violate Fodor's argument that modules have no external feedback path whereby the output of the module can affect its input.

5. THE STRUCTURE OF VISUAL SYSTEMS

In the field of machine vision, it has become commonplace to distinguish 'low-level vision' from 'high-level vision'. Basically, *low-level vision* takes the retinal input and codes it into a form suitable for interpretation of, or interaction with, the world but does not depend upon knowledge of what particular objects are in the world – though it may well depend on general properties of the world such as the hypothesis that the world is 'made up of relatively few connected regions' (section 3). By contrast, *high-level vision* is the process whereby this intermediate representation is used to guide the actual interpretation of the world in terms of objects or to determine patterns of interaction with, or navigation through, the world.

In the last two sections we have given some sense of what can be done in the way of layered computation in the early stages of visual processing. In this section, I will expand our analysis of modular decomposition by showing how the module for low-level vision may be decomposed into a small number of layered submodules, while high-level vision is decomposed into the interaction of a vast number of small modules called *schemas*. Such schema networks embody, though I shall not argue this explicitly here, many of the Quinean and isotropic properties (see section 6 for a definition) that Fodor holds to be a characteristic of central processes which distinguishes them from input modules.

It was J.J. Gibson who, perhaps more than anyone else, drew our attention to the immense amount of information about the structure of the world that could be inferred from the properties of surfaces and the way in which they reflected and transformed environmental energy as it passed to our receptors. However, the full impact of this work was delayed because the Gibsonians talked of 'direct perception', without addressing what to most of us would seem the self-evident fact that some mechanisms within the brain must be required to carry out the inference back from those environmental energies to the distal objects which transformed them. The sort of work reviewed in section 3 marked the beginning of using the insights of Gibson but embodying it in the computational framework that he himself rejected. David Marr, whom we have already mentioned, has been in the vanguard of those who would create a computational Gibsonianism, but his school has paid perhaps too little attention to the lessons that Gibson taught us. For example, Shimon Ullman, whose work on motion detection can be seen as giving computational expression to Gibsonian principles, in fact wrote a well-known article 'Against Direct Perception' which perhaps let the debates against Gibson's anti-computationalism obscure the debt to him.

In this section, I want to briefly look at the issue of what representations the low-level vision can deliver, and what methods are available for interpreting it. I shall contrast two overall specifications, one due to David Marr, and one due to Hanson and Riseman. In the upper half of Figure 6, we see that Marr postulates a one-way flow from the primal sketch (a fancy name for the sort of feature map that we have come to see as the result of the earlier stages of visual processing, as in the bug detectors in the retina of Lettvin et al., or the edge detectors in the cortex of Hubel and Wiesel). Marr then sees the interface between low-level vision and high-level vision given by the 2–1/2 D sketch which specifies for each portion of the visual field the depth of the corresponding distal object, and the local orientation of the surface at that point. This then provides the input to the object recognition process, and Marr and Nishihara (in a suggestion which was not implemented on the computer) have suggested that many objects can be represented in terms of the connection of a variety of generalized cylinders, and have suggested how the 2–1/2 D sketch might be processed to find such cylinders, whose description could be used to key a database of known objects.

Where Marr's theory is offered as a theory of the way the human visual system must work, Hanson and Riseman's work is offered as a useful way to build a machine vision system. However, I will draw

MARR

Object Recognition (not implemented)
↑
2–1/2 D Sketch*
↑
Primal Sketch*

HANSON and RISEMAN

Schema Assemblages (Implemented)
↑ ↓
Segmentation*
↑
Local Feature Extraction*

*Layered Computations

Fig. 6

lessons from both approaches for our understanding of the human mind. First, their system has a process for extracting local features, akin to the primal sketch, but since they are working with the recognition of images provided by single color photographs, their next level is not a depth map, but is rather a segmentation of the scene into regions demarcated from each other by such cues as color and texture (lower half of Figure 6). The resultant description of the image in terms of regions of various shapes and colors and textures then provides the input for high-level vision, which through a process of cooperative excitation of various schemas leads to the final interpretation of the image, through processes which I will outline later in this section. As will I think be clear from the discussion in section 4, it is not my intention to argue that Marr and Hanson are exclusively right in what they offer as the output map from low-level vision provides not one but several maps, which can be used perhaps to different ends, by processes involved in image understanding or the determination of behavior. I have suggested this in Figure 7 by showing that local organizing processes and aggregation procedures can provide a symbolic representation of the segmented image without a

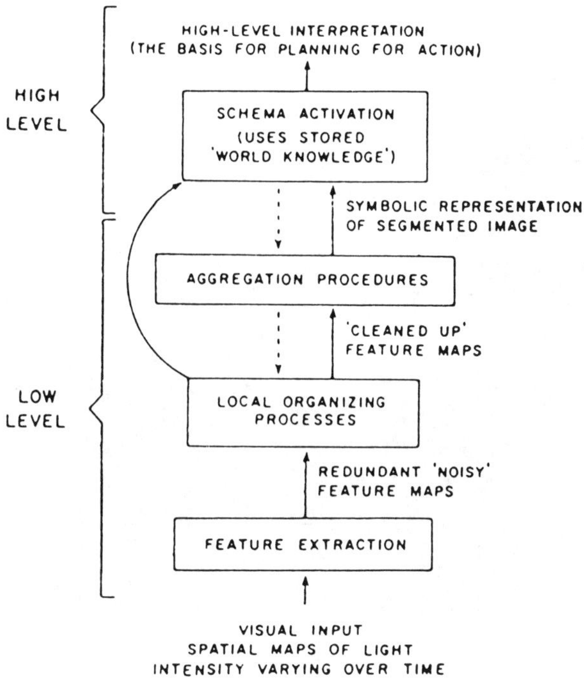

Fig. 7

picture is that low-level vision provides not one but several maps, which can be used perhaps to different ends, by processes involved in image understanding or the determination of behavior. I have suggested this in Figure 7 by showing that local organizing processes and aggregation procedures can provide a symbolic representation of the segmented image without a commitment as to what that representation might be. As I have said, I think that in fact it comprises several representations, and this certainly is born out by the increasing discovery of the division of the brain into 'many visual systems', as alluded to above. In the remainder of this section, I simply want to outline the approach that Hanson and Riseman have offered for high-level vision in terms of the interaction of schemas, since this approach embodies some of my own views about the sort of fine grain functional decomposition of the brain's activity that we are not yet able to follow through to detailed implemen-

Fig. 8A
A house image; and (B) the image segmented using low-level segmentation techniques.

tation in terms of neural networks. I see the reconciliation of the view of functional activity in terms of schema interaction with our growing, but still limited, knowledge of detailed neurophysiology as the major challenge for brain theory in the next fifteen years.

To complement the above general observations, we now present a more concrete discussion of how Hanson and Riseman orchestrate schemas in successful visual perception. Figure 8B shows the result of running segmentation algorithms on the image of Figure 8A. The top and bottom of the roof are fairly well delineated, but the left-hand edge is occluded while the right-hand edge 'bleeds' into the sky and foliage. The figure also illustrates that highlighting or variation in texture may lead the algorithm to subdivide a natural region into several segments, as we see for portions of the walls. The problem, then, is to design algorithms that can, on the one hand, take a region and split it into parts that are to be seen as giving us information about different objects, and

Fig. 8B

on the other hand, will aggregate regions that together characterize some distinctive portion of the image. The process of *image interpretation* calls on schemas which embody 'high-level' information about possible objects in the scene. For example, information about houses would, amongst other things, initiate a search for a near-parallelogram as candidate for the roof-image. However, the program would not fail if there were no parallelogram in the image, but might pursue more subtle possibilities, e.g. 'if you find two approximately parallel lines at top and bottom and portions of approximately parallel lines on the left and right, join up the lines, and explore the hypothesis that the resultant parallelogram is a roof'. Given a confident roof hypothesis, the system can hypothesize that below the roof the image will contain shutters or windows. Thus if regions there can be aggregated into a rectangle, the program can indeed follow the hypothesis that there is a window.

The *logic* of schema processing is inherently parallel. Each schema can have a separate instantiation, corresponding to each region for which there is a non-trivial confidence level for the object that the schema represents. All the active schema instantiations can then communicate with each other. In the fashion that we have seen before, the activity of a schema in one region may lead to the instantiation of a schema in another region to check context. If the newly activated schema does indeed find the expected context, then the confidence level of the original schema can be raised. Although research is under way on implementing schemas on computer networks, most work is currently done by simulating the interaction of schemas on a serial computer. Thus, a number of scheduling strategies may be imposed upon the logical parallelism of the schemas. The result of processing the image on Figure 8A with an early version of the VISIONS schema system is shown in Figure 8C. For further information on the processes involved, see Appendix 1 of Arbib[40] and Riseman and Hanson.[41]

High-level vision involves the interleaving, then, of multiple processes, a cooperative computation in which each is invoked where appropriate, possibly many times, with hypotheses being generated and discarded until the system converges on as good an interpretation as it is able to give with the facilities available to it. We claim that this style characterizes the perceptual mechanisms of brains, but it also is a useful model for 'central' processes.

6. THE PROSPECT FOR MODELLING CENTRAL PROCESSES

We may agree with the notion of a *module* within a cognitive system as a subsystem which has a well-defined set of inputs and a well-defined set of outputs, and with a well-defined relation between inputs and outputs. This relation may involve the mediation of state variables internal to the module but the values of these state variables will only have an effect upon other modules to the extent that they determine the output of the module. However, in the preceding sections we have sought to establish the following:

(a) Fodor's modules are only a subclass of modules in this sense, and his defining conditions may even be inconsistent.
(b) Fodor's modules are too big, in that he would view the visual system as a single module, whereas cognitive science must offer

Fig. 8C

an analysis in which the grain of the modules is at least as fine as that of the rectangles of Figure 3 or 7.

(c) In the light of (b), we reject Fodor's claim that there is no external feedback path whereby the output of the module can affect its input.

(d) The concept of module is not a new one initiated by Fodor, but has a long history both in functional analysis of cognitive systems and in the structural analysis of brain mechanisms of cognitive functions. The give-and-take between functional and structural (neural) decompositions of cognitive systems provides an important tool for cognitive science.

In this last section, I turn to Fodor's views on central processes. I could argue that the type of schema interactions posited for high-level vision in the previous section constitutes a viable model for the analysis of central processes, but I shall restrict myself to a critique of three dichotomies offered by Fodor as a grounding for his distinction between input processes and central processes. Rather disarmingly, Fodor admits that "there is practically no direct evidence, pro or con, on the question whether central systems are modular (. . .). When you run out of direct evidence, you might just as well try arguing from analogies, and that is what I propose to do ".[42] He assumes that "the typical function of central systems is the fixation of belief (perceptual or otherwise) by nondemonstrative inference. Central systems look at what the input systems deliver, and they look at what is in memory, and they use this information to constrain the computation of 'best hypotheses' about what the world is like ".[43] This leads Fodor to use scientific confirmation as the analogy to structure his model of central processes. The key to his argument is that scientific confirmation is:

> *Isotropic*: facts relevant to the confirmation of scientific hypotheses may be drawn from *anywhere* in the field of previously established truths, and
> *Quineian*: the degree of confirmation assigned to any given hypothesis is sensitive to properties of the entire belief system, e.g., based on a measure of simplicity, plausibility or conservation which is a metric over *global* properties of belief systems .[44]

Having accepted the force of his analogy, Fodor concludes that "Even if the flow of data is unconstrained within a module, encapsulation implies constraints upon the access of intramodular processes to extramodular information sources, [whereas] if (. . .) isotropic and Quineian considerations are especially pressing in determining the course of the computations that central systems perform, it should follow that these systems differ in their computational character from the [modules]" .[45] And on this basis Fodor claims [46] that the following three taxonomies are coextensive:

(a) FUNCTIONAL TAXONOMY : input analysis versus fixation of belief.
(b) TAXONOMY BY SUBJECT MATTER: domain specific versus domain neutral.

(c) TAXONOMY BY COMPUTATIONAL CHARACTER: encapsulated versus Quineian/isotropic.

He remarks that this co-extension, if it holds, is a *deep* fact about the structure of the mind. I shall argue not only that it does not hold, but that the individual taxonomies are not themselves useful.

(a) Input Analysis vs. Fixation of Belief. Fodor rejects Marr's primal, 2–1/2D and 3D sketches as defining the outputs of the visual processor, since then the visual input module would not provide for the recognition of objects and events. He argues [47] that basic perceptual categories constitute the output of the vision module – these are the 'middle levels' in implicational hierarchies, e.g., 'dog' rather than 'poodle' or 'thing'. In other words, it follows from Fodor's own account that the visual input module can, with appropriate visual stimulation, deliver to central processes a confident report that the distal stimulus is a dog. But in this case, no further central processing is required to fix the beliefs 'I see a dog'. Though it is incontestable that there are many beliefs whose fixation cannot be achieved by input analysis alone, Fodor's own view of the vision module denies that input analysis vs. fixation of belief constitutes a clear dichotomy.

(b) Domain-Specific vs. Domain Neutral. Well, this has been a useful taxonomy in AI, but Fodor's use of 'domain' is different, since 'vision' and 'language' are domains for him, but 'bacterial diagnosis' or 'the blocks world' apparently serve only as objects of ridicule. Fodor asserts[48] that "Early AI tried to treat central processes as though they were modular. Intellectual capacities were divided into (. . .) arbitrary subdepartments (. . .) and the attempted simulations proceeded by supplying machines with very large amounts of more or less disorganized, highly topic-specific facts and heuristics. (. . .) What emerged was a picture of the mind that looked embarrassingly like a Sears catalogue (. . .). I take it that the bankruptcy of this sort of AI is self-evident and constitutes a strong prima facie argument that the organization of central processes is not modular".

However, it is not self-evident to me that this approach is bankrupt. I think that many cognitive scientists would agree that it is only by using a 'microworld' to reduce problems of ambiguity and to limit the amount of information to be represented in a model that we can make the progress in analyzing process and representation which can provide the inductive base for the inference of more general mechanisms. However, a more telling point in the context of Fodor's argument is that scientific confirmation, his key analogue for central processes, itself 'looks like a Sears catalogue'. There are scientific societies for the study of physics, biology, chemistry, etc., and the volume of scientific research has grown to such an extent that the work of most scientists is confined to narrow subdisciplines with little communication between them: within cognitive science, work in linguistics is little affected by work in vision; within linguistics, a researcher in language acquisition may pay little heed to studies of historical phonology. We may decry this specialization, but it is neither 'embarrassing' nor 'self-evidence of bankruptcy'. If we accept AI usage, then the domain-specific vs. domain-neutral dichotomy does not match Fodor's other dichotomies, for an AI model of a central process will often span the dichotomy, using a domain-neutral 'inference engine' to process facts from a 'domain-specific' database – and many of us would find even this dichotomy to be mistaken, arguing that knowledge and processes are often intertwined. If we insist on Fodor's usage of domain, we find it is not defined, but is given by a list 'vision, language, . . .' whose continuation is unclear. Is 'reading' a domain, or is it to be regarded as 'sort-of-domain-specific', involving language and vision but not certain other domains, or is it 'domain-neutral' since neurological studies show it to involve many interacting submodules, including ones linked to vision, language and hearing? In any case, the dichotomy has become a trichotomy. But the situation is even worse than this. Lacking any clear definition of domain other than 'like an input module', Fodor's use of his dichotomy is vacuous, and any claim that it is a deep fact that 'domain-specific vs. domain-neutral' might match 'input module vs. central process' is reduced to empty tautology.

(c) Encapsulated vs. Quineian/Isotropic. Now that we have confronted the issue of scientific specialization, we see that this dichotomy does not usefully describe scientific confirmation. We may regard science as organized into modules with relatively restricted flows of data between them. To a first approximation, each specialty is encapsulated. However, work within the discipline is certainly Quineian and isotropic with respect to the facts and theories of the discipline itself, and only secondarily and with much longer time constants with respect to science as a whole. Since two can play at the analogies game, we might suggest that input modules correspond to scientific specialities while central

processes correspond to the integrative aspects of science. If we accept this analogy, encapsulation of the input processes is at best a first approximation; if we reject it, we suggest that modularity provides a good first approximation to the structure of the central processes. In either case, we reject the claim that the encapsulated vs. Quineian/isotropic dichotomy is coextensive with the input process vs. modular process dichotomy. Moreover, we may now counter the profound pessimism with which Fodor concludes his volume, when he argues[49] that "if central processes have the sort of properties that I have ascribed to them, then they are bad candidates for study". He ascribes what progress there has been in cognitive science to the fact that we have been able to study modular input systems. But if, as I argue, input processes are less encapsulated and some central processes are more encapsulated then Fodor believes, then there is no reason to accept the input process/central process taxonomy as setting a dispiriting limit to the success of cognitive science.

NOTES

* Preparation of this paper was supported in part by NIH NS14971. This is a revised version of a paper which has appeared in Jay L. Garfield (ed) (1987) *Modularity in Knowledge Representation and Natural-Language Understanding*, MIT Press.

1 M.E. Scheibel and A.B. Scheibel (1958) 'Structural Substrates for Integrative Patterns in the Brain Stem Reticular Core', *Reticular Formation of the Brain*, H.H. Jasper et al. (eds.), Little, Brown and Company, Boston, pp. 407–434.

2 W.L. Kilmer, W.S. McCulloch and J. Blum (1969) 'A Model of the Vertebrate Central Command System', *International Journal of Man-Machine Studies* **I**, pp. 279–309.

3 W.H. Pitts and W.S. McCulloch (1947) 'How We Know Universals, the Perception of Auditory and Visual Forms', *Bulletin of Mathematical Biophysics* **IX**, pp. 127–147.

4 J.Y. Lettvin, H. Maturana, W.S. McCulloch and W.H. Pitts (1959) 'What the Frog's Eye Tells the Frog's Brain', *Proceedings of the Institute of Radio Engineers 1959* **XLVII**, pp. 1940–1959.

5 T.P.S. Powell and V.B. Mountcastle (1959) 'Some Aspects of the Functional Organization of the Cortex of the Post Central Gyrus of the Monkey: A Correlation of Findings Obtained in a Single Unit Analysis with Cytoarchitecture', *Bulletin of Johns Hopkins Hospital* **CV**, pp. 133–162.

6 D.H. Hubel and T.N. Wiesel (1974) 'Sequence Regularity and Geometry of Orientation Columns in the Monkey Striate Cortex', *Journal of Comparative Neurology* **CLVIII**, pp. 267–294.

7 J. Szentágothai and M.A. Arbib (1974) 'Conceptual Models of Neural Organization', *Neuroscience Research Program Bulletin* **XII**, No. 3, pp. 310–479.

8 V.B. Mountcastle (1978) 'An Organizing Principle for Cerebral Function: The Unit

Module and the Distributed System', *The Mindful Brain*, G.M. Edelman and V.B. Mountcastle, MIT Press, Cambridge, Mass.
[9]J. Fodor (1983) *Modularity of Mind*, MIT, Cambridge, Mass., p. 13.
[10]*Ibid*. p. 40.
[11]*Ibid*. p. 21 and p. 37.
[12]*Ibid*. p. 21.
[13]*Ibid*. Section III. 5.
[14]*Ibid*. p. 41.
[15]*Ibid*. p. 42.
[16]*Ibid*. p. 45.
[17]*Ibid*. p. 101.
[18]*Ibid*. p. 102.
[19]*Ibid*. p. 103.
[20]*Ibid*. p. 20.
[21]M.A. Arbib (1972) *The Metaphorical Brain: An Introduction to Cybernetics as Artificial Intelligence and Brain Theory*, Wiley-Interscience, New York.
[22]M.A. Arbib and D.H. House (1985) 'Depth and Detours: An Essay on Visually-Guided Behavior', *Vision, Brain, and Cooperative Computation*, M.A. Arbib and A.R. Hanson (eds.), MIT, in press.
[23]M.A. Arbib, T. Iberall and D.M. Lyons (1985) 'Coordinated Control Programs for Movements of the Hand' *Experimental Brain Research Supplement* **X**, pp. 111–129.
[24]M.A. Arbib, E.J. Conklin and J.C. Hill (1986) *From Schema Theory to Language*, Oxford University Press.
[25]M.A. Arbib, C.C. Boylls and P. Dev (1974) 'Neural Models of Spatial Perception and the Control of Movement', *Cybernetics and Bionics*, W.D. Keidel, W. Handler and M. Spreng (eds.), Oldenbourg, Munich, pp. 216–231.
[26]P. Dev (1975) 'Perception of Depth Surfaces in Random-Dot Stereograms: A Neural Model', *International Journal of Man-Machine Studies* **VII**, pp. 141–202.
[27]S. Amari and M.A. Arbib (1977) 'Competition and Cooperation in Neural Nets', *Systems Neuroscience*, J. Metzler (ed.), Academic Press, New York, pp. 119–165.
[28]D. Marr and T. Poggio (1976) 'Cooperative Computation of Stereo Disparity', *Science* **CXCIV**, pp. 283–287.
[29]D. Marr (1982) *Vision: A Computational Investigation into the Human Representation and Processing of Visual Information*, W.H. Freeman and Co.
[30]D.H. House (1984) 'Neural Models of Depth Perception in Frog and Toad; Ph. D. Dissertation, Department of Computer and Information Science, Technical Report 82–16, University of Massachusetts at Amherst.
[31] See *op. cit.* (note 4).
[32] H. Barlow (1953) 'Summation and Inhibition in the Frog's Retina', *Journal of Physiology* **CXIX**, pp. 69–88.
[33] J.P. Ewert (1976) 'The Visual System of the Toad: Behavioral and Physiological Studies on a Pattern Recognition System', *The Amphibian Visual System: A Multidisciplinary Approach*, K. Fite (ed.), Academic Press, New York, pp. 141–202.
[34] J.P. Ewert and W. von Seelen (1974) 'Neurobiologie und System-Theorie eines visuellen Muster-Erkennungsmechanismus bei Kröten', *Kybernetic*, **XIV**, pp. 167–183.
[35] F. Cervantes-Perez, R. Lara and M.A. Arbib (1985) 'A Neural Model of Interactions

Subserving Prey-Predator Discrimination and Size Preference in Anurans', *Journal of Theoretical Biology* **CXIII**, pp. 117–152.
[36] R. Lara, M.A. Arbib and A.S. Cromarty (1982) 'The Role of the Tectal Column in Facilitation of Amphibian Prey-Catching Behavior: A Neural Model', *Journal of Neuroscience* **II**, pp. 521–530.
[37] G. Székely and G. Lázár (1976) 'Cellular and Synaptic Architecture of the Optic Tectum', *Frog Neurobiology*, R. Llinás and W. Precht (eds.), Springer-Verlag, pp. 407–434.
[38] See *op. cit.* (note 35). See also M.A. Arbib (1981) 'Perceptual Structures and Distributed Motor Control', *Handbook of Physiology – The Nervous System. II. Motor Control.* V.B. Brooks (ed.) American Physiological Society, pp. 1449–1480.
[39] M.A. Arbib (1987) 'Levels of Modelling of Neural Mechanism Underlying Visuomotor Coordination', *Behavioural and Brain Sciences*, **X**, pp. 407–465.
[40] See *op. cit.* (note 39).
[41] E.M. Riseman and A.R. Hanson (1987) 'A Methodology for the Development of General Knowledge-Based Systems', *Vision, Brain and Cooperative Computation*, M.A. Arbib and A.R. Hanson (eds.), MIT, Mass., pp. 285–328.
[42] See *op. cit.* (note 9), p. 104.
[43] *Ibid.* p. 104.
[44] *Ibid.* pp. 107–8.
[45] *Ibid.* pp. 110–111.
[46] *Ibid.* p. 112.
[47] *Ibid.* pp. 94–5.
[48] *Ibid.* p. 103; p. 139 note 43.
[49] *Ibid.* p. 137.

Departments of Computer Science and Neurobiology
University of Southern California, Los Angeles, U.S.A.

HUW PRICE

ACTION EXPLANATION AND THE NATURE OF MIND*

1. INTRODUCTION

Action explanation has long been seen as a guide to psychological theory. The classic argument is Hume's. Hume's distinction between reason and the passions rests on the principle that reason is motivationally inert: "The reason alone can never be a motive to any action of the will; (. . .) it can never oppose passion in the direction of the will".[1] So to explain an action fully, Hume thinks, we must always appeal not only to an agent's 'reason', or beliefs, but also to their 'passions', desires, or 'pro-attitudes'.[2]

Given that we want our psychological theory to be capable of explaining action, the principle that belief is motivationally inert thus puts a fundamental constraint on our view of the gross architecture of the mind. This principle is as orthodox today as it was controversial for Hume. As a result, despite its distance in other respects from Hume's view of the mind, contemporary cognitive theory takes his 'two-box' psychology virtually for granted. (By way of corollary, Hume's principle continues to be applied in areas of philosophy that depend on this psychology. In particular it is still taken, as Hume himself took it, to provide a foundation for various brands of ethical non-cognitivism. Briefly, the non-cognitivist wants to distinguish moral commitments from genuine beliefs; given Hume's principle, it is sufficient to show that such commitments are active, rather than inert. More on this in section 5.)

I am going to argue that this contemporary reliance on Hume's principle is unjustified. The most charitable contemporary reconstructions of Hume's arguments fail to show that beliefs alone can never motivate intentional action. As a result they fail to show either that action explanation requires a two-box cognitive theory; or that moral commitments (even if agreed to motivate actions) are not beliefs.

However, my approach will be rather indirect. It seems to me that there is a similarity between Hume's argument and some much more recent attempts to show that action explanation puts constraints on the nature of mind: arguments to the effect that action depends on demonstrative or indexical beliefs. The similarity seems to me to be of consid-

Peter Slezak and W.R. Albury (eds.), Computers, Brains and Minds, 221–251

erable interest in its own right, as well as a good guide to the fundamental weaknesses of Hume's argument. So I shall first consider the demonstrative and indexical arguments, and try to show that they are invalid, at least as interpreted as constraints on arbitrary cognitive systems. I shall then argue that contemporary arguments for Hume's principle rely on similar moves, and hence fail for much the same reasons. The detour will be justified, I hope, by the interest of the scenery along the route.

2. ACTION AND INDEXICAL THOUGHT

In a paper entitled 'Demonstrative Thought and Psychological Explanation'[3] Christopher Peacocke argues for what he calls the *Indispensability Thesis*:

IT: No set of attitudes gives a satisfactory psychological explanation of a person's acting on a given object unless the content of those attitudes includes a demonstrative mode of presentation of the object.[4]

As it stands, this is a claim about persons, rather than about arbitrary cognitive systems. Even so, it admits of more than one interpretation: Is it a claim about what the minds of human agents actually do contain, or about what they could conceivably contain, given their organic limitations? If the latter, is there anything to prevent us generalizing to arbitrary cognitive systems – does the reference to 'a person' do any work?

The argument Peacocke actually offers for IT is not much help here. It consists largely in examples: in cases in which it seems intuitively clear that an action explanation does require a psychological attitude with demonstrative content. Such examples may convince us that certain human actions are in fact inexplicable without appeal to demonstrative thoughts: but it seems unlikely that they could show that the same would be true of the actions of any conceivable cognitive system, or even of any human agent. However, we shall see that Peacocke's examples do suggest arguments for a more general conclusion – arguments for regarding IT as a constraint on arbitrary cognitive systems.[5]

A similar conclusion to Peacocke's, though concerning indexical rather than demonstrative beliefs, seems to have been widely drawn from John Perry's influential paper, 'The Problem of the Essential Indexical'.[6] Perry argues that in order to explain why an action is performed *when*, *where* and *by whom* it is, we need in general to ascribe an indexical belief to the agent concerned. Like Peacocke, Perry argues by example, describing cases in which it seems clear that without indexical beliefs, we do not have an adequate explanation of a particular action. Again such examples show at best that certain human actions are in fact inexplicable without indexical beliefs, and not that the same would be true of the actions of any human agent, let alone of any conceivable cognitive system. This may be all that Perry himself requires. But again his examples suggest an argument for a more general conclusion (and have, I think, been taken to do so).

The most plausible and distinctive of these arguments runs something like this. An action is an intentional change in a person's (or cognitive system's) behaviour. To explain it we need to explain why the change happened when, where and in whom it did happen. This requires a change in the agent's cognitive apparatus (hereafter, 'Cog'): the action is a result of something that happens in Cog, and uniformity there cannot explain a difference elsewhere. But in general it is only the agent's indexical belief states that do change in the required fashion. For example I may have believed for weeks that I was to start reading this paper at 11 o'clock on the third Tuesday in May. That belief did not change at the moment I started. What did change in me, and hence what explains my starting, was that I acquired the indexical belief that it was *now* 11 o'clock on the third Tuesday in May. That indexical belief thus played a crucial role in the production of my action. It filled an otherwise mysterious gap in the chain of psychological events, or states, that led me to open my mouth.

I shall call this the *Causal Role Argument*. With some qualms, but for the sake of argument, I shall attribute it to Perry. Since the argument applies as much to possible robots as to actual humans, I thus take Perry to be arguing that indexical beliefs are essential in principle, as well as in practice.[7]

Perry treats demonstratives as a kind of indexical. In one of his examples an agent has to believe 'that *this* is the Mt. Tallac trail and *that* is Gilmore Lake'.[8] So Peacocke's conclusion would seem a special case of Perry's. Whether the inclusion is proper will depend on what Peacocke means by an object (as of course will the scope of Peacocke's argument in general). However, if times, places and people are all to be treated as objects, then the two conclusions might be expected to be co-extensional.

However, this ignores what might be thought a more significant difference between the two conclusions: that they seem to focus on different aspects of the explanation of action. Perry seems concerned with the causal role of an agent's awareness of a time, place, person or other object, in the production of an intentional action; Peacocke with the 'directedness' of an action towards a particular object. The Causal Role Argument seemed to rest on the intuition that an intentional action is an effect, *inter alia*, of certain features of an agent's environment. Since the immediate causal history of an action lies in Cog, relevant features of an agent's environment must be represented in Cog, in order to explain intentional behaviour. Perry's conclusion then rests on examples that suggest that only indexical belief states are appropriately sensitive to certain environmental features. Peacocke's conclusion, however, involves the notion of 'an agent's acting on a given object'. This suggests a concern with the effects of an action, rather than its causes. Perhaps more importantly, it seems to involve the idea that these effects may be *directed at* or *targetted on* particular objects. The problem is to cash this directedness in terms of the contents of Cog (and the claim is that only demonstrative thoughts will do the trick).

I think this difference will turn out to be fairly superficial. If an action is directed towards a given object in an agent's environment, that object presumably plays a causal role in the production of the action. If the object had not been there, the agent would not have acted in that way. Conversely, any feature of an agent's environment which, via a representation in Cog, plays a causal role in the production of an action, has a claim to be a feature towards which that action was 'directed'. Unless directedness amounts to more than an agent's intentional involvement of one object rather than another, we would seem justified in speaking of actions as being directed towards times and places, as well as to more concrete objects.

At present, however, there is nothing to be gained by insisting on assimilating the two conclusions. In the next section I shall therefore examine possible arguments for Peacocke's IT. The following section will then look at Perry's argument; and show, I think, that in so far as it rests on different grounds, it still does no better than Peacocke's.

3. IS 'THAT' REALLY NECESSARY?

Peacocke's main example involves an agent who reaches for a container that he believes to contain life-saving pills. We want to explain why the agent reached for that particular container. It seems we need to say that he believed that it was *that* container that contained the pills; or at least that it was *that* container which met some other description (such as 'left by the doctor') that he took to entail that it contained the pills.

> The thought behind the Indispensability Thesis is that however many descriptive modes of presentation 'the Φ' with which we supply the agent for his thoughts 'the Φ contains the pills I need', we will not supply attitudes sufficient to explain his acting on the container in front of him unless the agent has a thought with the same predicate 'contains the pills I need' in which that container is presented demonstratively.[9]

The question here is: Why did the man reach for *that* object? A first thought might be that IT is somehow a product of the way in which we have framed the question – of the fact that in characterizing the action to be explained, we have ourselves referred to the object demonstratively. We can check this by substituting some other mode of presentation of the object concerned. Why, for example, did the agent reach for the bottle that used to belong to Uncle Fred? Again, because he believed that it – the bottle that belonged to Fred – contained the life-saving pills (and he wanted to live). Does this mean that object must have been presented to the agent as the bottle that belonged to Fred? Clearly not: in the last sentence but one the referring expression is used transparently. The agent believed of what we can refer to as 'the bottle that belonged to Fred' that it contained the pills; but he himself need not have referred to it this way.

Why then should we not say that in agreeing that the agent believed that *that* contained the pills, we mean the demonstrative to be taken transparently? And hence that the agent need not have thought of the bottle demonstratively? The answer seems in part that we take for granted that the agent would have thought of it demonstratively – or at least would have had access to such a mode of presentation. He might not have known it as the bottle that used to belong to Fred, but he must have been aware of it as *that bottle* (or perhaps simply as *that thing*).

In other words, IT might seem to depend on what we might call the *Availability Assumption*:

AA: An agent who is acquainted with an object in such a way as to make it a possible target of an intentional action, is also sufficiently acquainted with it to refer to it demonstratively.

This principle has some plausibility, I think. An object sufficiently differentiated from its surroundings to be the target of an intentional action, seems sufficiently differentiated to allow demonstrative reference. However, it is not clear that AA guarantees IT. It is one thing to have access to a mode of presentation of an object; another for that mode of presentation to play an indispensable role in psychological explanation. It seems to me that there are two possible arguments for the extra step. I shall call them the *Argument from Doubt* and the *Targetting Argument*.

The Argument from Doubt relies on the intuition that (perhaps in virtue of AA) demonstratives provide a kind of standard, in terms of which it is held to be possible to establish the referents of other referring expressions. Thus in general demonstrative expressions enable us to enquire about – and more significantly, to *doubt* – the designations of other expressions. 'Is *that* George (the last bus, the price of fish)?', we can ask, when we don't know what 'George' ('the last bus', 'the price of fish') refers to; or, 'Is that *really* George (the last bus, the price of fish)?', when we doubt it.

This doubt-enabling feature of demonstrative modes of presentation provides an argument for Peacocke's thesis, along the following lines. The man who reached for the pills could have doubted whether *that* container was the one that contained the life-saving pills. In the grip of such doubt, he would not have reached for it (or at least his doing so would not be explained by the beliefs and desires we are presently invoking). So to explain his action we need to assume that he did not have such a doubt – in other words, it seems, that he believed that that container did contain the pills. In other words, he must have thought of the container under a demonstrative mode of presentation.

Note here the use of the assumption that a demonstrative mode of presentation of the container is available to the agent. We might try to argue (i) that if the agent had doubted that the bottle that used to belong to Uncle Fred contained the pills, he would not have acted as he did; and therefore (ii) that explanation of his action must invoke a thought in which the container is presented to him as the bottle that used to belong to Fred. However, this version of the argument trades on an equivocation between a transparent and an opaque use of the referring expression 'the bottle that used to belong to Fred'. (i) is only plausible under a transparent reading; whereas (ii) requires that (i) be justified as an opaque attitude ascription. We have no reason to think that the agent knows the container under this description; and hence none for thinking that the belief he would express by 'The bottle that used to belong to Fred contains the pills' is a precondition of his action.

Thus in the demonstrative case AA saves the Argument from Doubt from what would otherwise be a fallacy of equivocation. The argument fails for other reasons, however. For one thing, it relies on the assumption that the lack of a disbelief entails possession of the corresponding belief – in other words that we believe whatever we do not doubt. In the case considered, this assumption is required to support the inference from the negative claim that the agent concerned did not doubt that *that* bottle contained the pills, to the positive conclusion that he believed that that bottle did contain the pills.

Do we believe everything that we do not doubt? In general, clearly not. Many propositions are undoubted in virtue simply of never having been entertained. To say that in such a case we believe whatever we do not doubt would be to ascribe contradictory beliefs to ourselves – for in such cases the negations of the propositions concerned have also not been entertained. The general principle needs to be qualified. We want to allow for unconscious beliefs – cases in which the lack of a pattern of behaviour that would reveal a doubt, is evidence of an implicit belief. But we don't want to say that lack of any doubt whatsoever is equivalent to possession of a corresponding belief.

On any account, a crucial factor will presumably be the availability of circumstances in which a doubt could be manifest. Lack of manifestation of a doubt is to be taken to be manifestation of the corresponding belief. In this way the ascription of the belief will be given a behavioural justification. However, a lot will hang on the interpretation of the phrase 'in which a doubt could be manifest'. For example, we might say that every time I eat an apple, I have the opportunity to manifest a doubt that it has not been poisoned. But it would be absurd to conclude that on every such occasion I believe that my apple has not been poisoned. If this were the principle of belief ascription that underlay IT, then satisfactory psychological explanation of action would be impossible. For there would be no end to the number of such beliefs we would have to ascribe, to explain any action whatsoever.

This points to a second and more basic flaw in the Argument from Doubt. Not only is the lack of a doubt not equivalent to a belief; but the explanation does not have to mention that lack – on pain, clearly, of having to mention the lack of endless doubts. The principle we need to preserve is this: an action A may be explained by attitudes $P_1, P_2, \ldots P_n$, even though it would not be explained by – indeed, would not have been performed given – attitudes $P_1, P_2, \ldots P_n$ and R. There should be nothing surprising about this principle. It applies to causal explanation in general. Causes are usually at best sufficient *in the circumstances* for their effects. And with this principle in place, the Argument from Doubt is inadequate to support IT.

It is unlikely that Peacocke intends IT to rest on the Argument from Doubt. Despite its deficiencies, however, the argument does suggest a more subtle argument for IT – for the two arguments share a common assumption.

Consider again Peacocke's hypochondriac. Suppose now that like us, he believes that the pill bottle in question used to belong to Uncle Fred. 'Why did you reach for the bottle that used to belong to Fred?', we ask him. 'I reached for the bottle that contained the pills', he answers; 'It happened to be the one that belonged to Fred'. In this case it is clear that although in fact he believed he was reaching for the bottle that used to belong to Fred, this belief plays no role in explaining his action. If he had doubted the same thing, it would not have made a difference.

If analogous reasoning applied to the demonstrative case, it would defeat the Argument from Doubt. Why is it then that when we ask the hypochondriac why he reached for *that* bottle, we should not be satisfied with the explanation: 'I reached for the bottle containing the pills; it happened to be *that* one'?

The answer seems to be something like this: in order to direct an action towards a particular object, we rely on our perceptual access to the object in question. Targetted action requires perceptual guidance: an action cannot be directed at an object, unless that object is distinguished from other objects (and from any 'background') in an agent's perceptual field. Moreover, if Cog is to control a targetted action, the target object must not only be represented in Cog; its representation must be suitably related to the object's distinguishing perceptual characteristics.

The point can be illustrated by a series of simple examples. At one extreme, consider a hunter who sets passive traps for some sort of prey:

a web-building spider, for example. When a spider catches a fly in this way, we would not say that the spider's action was directed at that particular fly. On the contrary, the fly just happened to satisfy the various general parameters – size, spatiotemporal location, and so on – in virtue of which any object would be captured by the web. As in the case of a pond that is hit by a meteor, to use Peacocke's example,[10] the fact that it is *that* fly is a kind of accident of history.

Next consider an intermediate case. Like spiders, many frogs make a living catching flies. But they usually do it with their tongues, a feat that presumably requires perceptual control. The frog's perceptual apparatus guides its tongue to a passing fly. So there is a sense in which this action is targetted on a particular object. The difference shows up in counterfactuals: if the fly had been in a slightly different place, the frog would still have caught it; but if the spider's fly had flown a little to the left, it would have missed the web. (Of course there are limits: if the fly had been in a different swamp altogether, even the frog would not have caught it.)

However, the frog's action is still not *cognitively* targetted. This is not primarily because the frog lacks a cognitive unit. Even if frogs were as intelligent as we are, their cognitive control of their tongues (*qua* fly-catchers, at any rate) could be expected to be limited to stop/start instructions. Feeling peckish, the frog's Cog would thus issue the instruction: Catch a fly! Not 'Catch *that* fly!', for the perceptual system that guides the tongue to the fly just isn't linked to Cog. Perhaps it used to be, but those frogs weren't so good at catching flies – by the time they had finished cogitating, the fly had usually flown. So the intentional act of fly acquisition is better made non-specific – any fly will do.[11] This means that although the bodily movement is perceptually targetted to a particular fly, from Cog's point of view it is still an accident of history that a particular fly turns up for lunch.

Full cognitive targetting thus seems to require continuous cognitive control of a perceptual guidance unit. The crucial point – the crux of the Targetting Argument – is that in such a case the cognitive mode of presentation of the targetted object must be appropriately linked to the perceptual distinctions employed by the perceptually-guided targetting unit. Cog can be said to control a targetted action only if several conditions are satisfied. For a start, Cog must be capable of discriminating the object concerned from its background to the same extent as the relevant perceptual apparatus (i.e., as that part of the agent's perceptual apparatus that is linked to the relevant motor systems). But mere discrimination isn't enough – a frog Cog might be capable of following the fly-catching operation, without controlling it. Cognitive control requires a suitable flow of information between Cog and the perceptually-guided targetting unit. It would seem that Cog must be capable of representing the target object in perceptually-grounded terms. Cog must make a selection from a range of alternatives presented by perception – in effect, it must point to something in its perceptual field. Otherwise it will once again be as if Cog had issued the *general* instruction: Find something – anything – that matches the following description: . . . .

At this point demonstrative modes of presentation seem just the job. Demonstrative reference seems perceptually-grounded. Demonstratives seem to work by calling attention to distinctive features of a person's perceptual field. In principle any perceptual distinction seems a possible basis for a such a mode of presentation – and in particular, any distinction exploited by a perceptual guidance system. Demonstratives thus seem to supply the right kind of link between Cog and perception. And on this basis the Targetting Argument claims that unless an object is picked out for an agent by a demonstrative mode of presentation, then an action cannot be considered to be intentionally directed at that particular object.

However, it is one thing to identify a function that demonstratives are well-suited to serve; another to show that only demonstratives will do the job. I am going to argue that the required action-guiding link can be (and sometimes is) established in other ways; and hence that IT does fail in certain (admittedly exceptional) cases, and that it could fail much more widely. I begin with an example which, though perhaps not a counter-example (it depends on what we mean by an object), at least suggests where we should look for counter-examples. The example involves the action of choosing one's words, of deciding to express oneself in this way rather than that.

G describes *F* as 'Kerr's cur'. An unusual phrase – why did *G* choose it? We can imagine that before his mind's eye, as he stepped forward to speak, *G* could see a number of fitting ways of describing *F*. Why did he choose that one? Clearly he wished to describe *F* in derogatory terms. But he would be speaking in public, indeed to history. The occasion called for a memorable phrase. Given *F*'s association with a man named Kerr, 'Kerr's cur' must have seemed (as *G* put it, years later) 'very apt'.

Do we need to add that *G* believed that *that* phrase was the phrase

'Kerr's cur'? Apparently not, so long as we avoid a type-token confusion. What needs to be explained is *G*'s choice of an expression of a certain *type* (from the range of epithets that occurred to him as he reached for the microphone). His use of a particular *token* is in general fully explained by his choice of the type. True, there could be cases in which we would need to invoke *G*'s belief that *that* phrase – the noise that came from his lips – was a token of a certain phrase-type. But these are irrelevant to the present case, in which what we want to explain is the choice of the phrase-type in the first place.

What happens as *G* makes his choice? Let us imagine it is something like this: he mentally surveys a shortlist of candidates (already selected in some other way). Having made his choice, he loads the chosen phrase into a appropriate speech register. The question is: In loading the chosen phrase, does *G* have to think of it as 'that phrase'? It seems not – and the reason seems to lie in the similarity between on the one hand the phrases themselves, as they appear on *G*'s mental list, and on the other their modes of presentation in Cog. For it is this connection that would make it absurd to object that *G* chose, as it were, only the description of a phrase, and that the actual phrase he used just happened to be the one that matched this description. The relationship between words and their names is no mere accident.

It might seem that the example hangs on the fact that all the action takes place in *G*'s head – and that it is only in a metaphorical sense that *G* can be said to 'choose' or 'pick out' the phrase 'Kerr's cur'. But suppose that *G* actually had a written list of suitable phrases (his speech writer having thoughtfully jotted down a few suggestions). Before he speaks, *G* runs through the list, and draws a circle around 'Kerr's cur', to remind himself to use it. His circling the phrase is presumably a perceptually-guided action. What instruction did he issue to his perceptual guidance system? Was it 'Circle *that* one', or simply 'Circle "Kerr's cur"'? The important thing is that the latter would do. In effect, it presents the perceptual guidance system with something that it can match against its own perceptual information – a picture of the targetted object. (The instruction is: Find something that matches 'Kerr's cur', and then circle it.) So there is no longer an essential role for the demonstrative thought. Of course, *G* could issue a demonstrative instruction, but he doesn't have to.

All the same, it may seem that no objects more concrete than words can be targetted non-demonstratively; and hence that IT holds, so long as suitably restricted to 'proper' physical objects. These days we are well aware that there are few interesting similarities between names and the objects they refer to. Even onomatopoeia is rather uncommon. Names are normally the products of quite arbitrary linguistic conventions. In other words, the representations of objects in Cog are likely to bear no intrinsic relation to the perceptual images of the objects they represent.[12] Only in rare cases, such as when the referent is itself a word, are linguistic representations anything more than arbitrary symbols. Language is generally non-perceptual (or should we say that perception is generally non-linguistic?).

However, we should avoid the fallacy of unwarranted generalization. After all, we are more interested in principled constraints on arbitrary cognitive systems than in contingent human limitations. The problem, we have seen, is to secure an appropriate action-guiding link between cognition and perception – roughly speaking, to provide a flow of information between the two, in a form usable in cognition, and of such a content as to allow perceptual guidance of an action towards a particular object. And the difficulty seems to stem from the fact that there is generally no natural similarity between the representation of information in perception and its more symbolic repesentation in cognition. So it is as if we are required to link two different transport systems: the link has to match one to the requirements of the other. Let us consider the range of possible solutions to such a problem.

The transport analogy suggests two basic solutions: from a given base, either system can be extended to meet the other. Cognition can be extended to employ as linguistic symbols, or forms of representation, the existing outputs from perception. Or perception can be extended to supply outputs in a form already appropriate to the symbolic capabilities of cognition. We shall see that both strategies are capable of providing non-demonstrative targetting of objects of many kinds. In principle these strategies could be combined, producing a wide range of hybrid solutions. In transport terms this would correpond to extending both systems in order that they should meet in the middle.

There is also third basic strategy, that I shall not explore in any detail. It corresponds to the construction of a bridging system, matching the output of one transport network to the input of the other. At least one implementation of this strategy would exploit the role of 'similarity' in connecting cognitive and perceptual symbolism. The strategies we are going to discuss will turn on modifications to cognition, perception, or both, which would have the effect of producing such similarities. But the notion of innate or objective similarity is highly problematic. Similarity often lies in the eye of the beholder. This suggests that the required link might be achieved by modifications in our 'similarity sensors'. In other words, the required associations between cognitive and perceptual representations might simply be innate. We might be built to perceive a 'natural' connection between redness and the word 'red'. In what follows, however, I shall rely on our existing intuitions about the similarities between modes of representation.

The first strategy involves the use of symbols that are not conventional words, and whose meanings depend on (what we take to be) their natural similarities to our perceptual images of the things they refer to. This sort of thing is actually quite common. In speech, we often refer to a sound simply by mimicking it. And in writing, we quite often introduce visual representations, or likenesses, of something to which we want to refer. Here are a couple of examples, one written and one meant to be spoken[13]:

(i) The child grabbed the steering wheel and said [brooom, brooom] (here I imitate a child's imitation of an engine noise).
(ii) Someone has painted ☮ on the wall of our embassy.

It is not difficult to see that as in the 'cur' case, explanation of the actions described in these examples might well have no need of a demonstrative mode of presentation of the noise or the peace symbol. For example, we can say that the child wanted to make a noise like a car, and believed that [brooom, brooom] is a such a noise. It would be pointless to add that she believed that *that* noise (i.e. the noise that she made) was the noise [brooom, brooom].

Clearly the number of such cases will depend on what sorts of things can be used as cognitive symbols. A language based on pictorial images, for example, would enable an agent to target ordinary physical objects. In a way this is what demonstratives provide. They direct us to something in our perceptual representation of the context of utterance, enabling that context to play a role in determining the meaning of what is said. The advantage of such a system lies in its extraordinary generality, or universality. It enables us to generate a symbol for virtually anything in our physical environments, provided only that it is distinctive enough to stand out in perception to ourselves and our fellow speakers. And it thus provides linguistic access to an environment whose diversity would otherwise far outstrip the symbolic resources of human language.

However, demonstrative reference achieves its universality at some cost. In communication it depends on the existence of a shared perceptual framework. And it has the usual disadvantage of a multi-purpose tool: to all intents and purposes, it can only do one thing at once. It is quite conceivable that particular cognitive systems, in particular circumstances, would be better served by more specialized linguistic tools: for example, by the ability to use pictorial images themselves as linguistic symbols. This would differ from the demonstrative case in allowing the image itself to function as a symbol, rather than employing a surrogate symbol, or 'pointer', whose function is to call attention to something in the perceptual field.

Consider for example the cruise missile, which is said to employ pattern-matching techniques in locating its target. Its basic instruction takes the form 'Go to [picture]'. The missile is designed to fly to something that looks like [picture] (and then explode). It is true (one hopes) that existing cruise missiles are no more than perceptually-guided motor units – their instructions come from outside. But more advanced models might choose their own targets, and give themselves instructions of the form 'Go to [picture]'.

So much for extending the symbolic repertoire of the cognitive system. The second approach is to enhance the perceptual apparatus, in order to make perceptual information available in Cog's existing symbolic format. An example in the same vein as the last: bomber pilots used to depend on demonstrative presentations of their targets. 'Das ist London', the Luftwaffe presumably thought, as they steered towards the glow in the western sky. Modern technology has made such thoughts redundant. Modern military aircraft carry navigational computers, linked to a device to project a three-dimensional image in the pilot's visual field. The computers calculate the aircraft's position in relation to its target, and might for example project two small circles, which appear to the pilot superimposed on the visible landscape. One circle appears over the target, and is labelled 'TARGET'. The other appears at infinity in the aircraft's present direction, and is labelled 'PRESENT DESTINATION'. To reach the target, the pilot need only steer the aircraft so as to keep the two circles concentric. The pilot is still a human agent, capable of directing intentional action towards objects in the external world. The difference is simply that thanks to the enhanced perceptual system supplied by the aircraft, IT fails. Some of the pilot's perceptually-guided intentional actions do not depend on demonstrative presentation of his or her environment.

Suppose for example that the target is a pond, and the pilot successfully bombs it, having thought of it simply as 'TARGET' (the description used by the enhanced perceptual system provided by the aircraft). Is the fact that that particular object was bombed a kind of accident, reflecting the fact that it 'just happened' to be the object that matched that description? No, for the description is perceptually linked. Even if the pond had been somewhere else the description would still have applied to it – which means that *in virtue of the fact that the corresponding cognitive symbol would have had a different reference*, the pond would still have been bombed.

At this point it may be objected that if the computer had picked out something else as the target, the pilot would have directed both action and aircraft towards some other object. Doesn't this mean that the fact that the pilot bombed a particular object is not explained by the pilot's cognitive state – but only by the conjunction of this and the fact that the computer picked out the object in question? I shall return to this objection in a moment. (In fact there is nothing here not already a product of ordinary perceptual systems.)

The bomber example trades on the fact that in their professional capacity, at any rate, bomber pilots are called on to perform a very limited range of actions, on a rather limited range of objects. In other words, theirs is just the sort of environment in which demonstratives might be expected to be dispensable. For we saw that demonstratives are multi-purpose names – in effect, a single universal name – and the value of a multi-purpose tool is proportional to the number of tasks at hand. In an effectively finite and perceptually well-discriminated environment, a cognitive system may well do better with individual perceptually-grounded names.

One more example, to illustrate this point: consider the central taxation computer. It inhabits a universe of taxpayers, perhaps divided into a number of natural kinds: individuals, corporations, charitable bodies, and so on. The computer knows each entity in this universe of taxpayers by its own unique name. All the information the computer receives about its universe is expressed in terms of these names; as are all the instructions it issues concerning particular taxpayers. There is simply no place for demonstrative thoughts.

Does this mean that the computer is incapable of directed action, of action targetted on a particular taxpayer? In favour of this view, it may be said that to explain the fact that the computer acts on a particular individual – issues me with a demand for a billion dollars in back taxes, for example – we need to add to our account of the computer's cognitive state, at least the information that I am what the tax office refers to as 512/879/3274. The computer merely instructs its subordinates to send 512/879/3274 a demand for a billion dollars. To explain the effect of this instruction, we then have to look outside the computer, and tell the causal story in virtue of which the letter in question ends up in my mailbox.

This is the objection we deferred in the previous example. There, the corresponding claim was that to explain why the pilot bombed a particular object, we have to note that the aircraft's computer picked that object out as the target, rather than something else. The objection threatens the entire strategy of perceptual enhancement. An enhanced perceptual system, it suggests, would leave one curiously out of touch with the world – so much so that to explain one's actions, we would need to refer to the mediating perceptual system. Indeed, it is easy to see which slippery slope the objector thinks we are on: such a perceptual system might simply be *deceiving* the cognitive system that possessed it. The pilot might simply be operating a rather realistic flight simulator – while the unfortunate tax computer might simply be sitting on a test bench, the electronic equivalent of a brain in a vat.

In fact, however, this slope slopes the other way. The appeal to the possibility of illusion backfires, because in so far as it is problem at all, it affects ordinary perceptually-grounded explanations as much as these imaginary cases. Suppose for example that among his other problems, Peacocke's hypochondriac suffers from hallucinations. He sees bottles of pills the way desert travellers see oases. When at last he grasps the real bottle, we need to add to our explanation a note to the effect that in this case there was a bottle where he thought he saw one. (Whether we regarded this as a kind of accident, or whether we said that he had actually perceived the bottle, would presumably depend on the details of his condition.)

Ordinarily we take for granted that other people's perceptual systems function reliably, and much as our own. Without this tacit assumption, demonstrative communication would be impossible. With it, we have the Availability Assumption, and a part of the background of ordinary action explanations, which determines what we take to call for explanation. In effect, it means that action explanations are not usually required to unpack demonstrative references. But it does not mean that there is nothing to unpack, no causal story to be told about the role of demonstrative thought, perception and the world in the production of action. There is a story, as there is in our perceptual enhancement examples. Whether an explanation has to tell it depends in every case on what we are prepared, or allowed, to take for granted. Demonstrative thoughts have no favoured status here – or at any rate none sufficient for Peacocke's Indispensability Thesis. Where perception does not discriminate between objects, not even demonstrative thoughts can explain the fact that we act on one rather than the other.

The Targetting Argument thus offers a promising account of the special role that demonstrative thoughts seem to play in action explanation. But at the same time it leads us to a variety of actual and possible counterexamples to the Indispensability Thesis. IT seems to be false in all its various interpretations, and certainly so in its strongest and most interesting ones. In the general case, directed intentional action does not seem to depend on demonstrative thought. Demonstratives merely provide one of a number of conceivable solutions the problem of cognitive representation of the possible objects of intentional action. The problem has several variables: the size and variability of the universe of action, the symbolic resources of cognition, and the nature of the perceptual system. In general the advantages of demonstrative modes of presentation will depend on the balance of these factors. In particular cases – perhaps in our own, most of the time – they may indeed provide the only solution. But if and when this is true, it is clearly due to the peculiarities of the case in question, and not to any general need for demonstrative thoughts.

4. I, ROBOT?

I think that a similar strategy can be used against the Causal Role Argument, which in section 2 I attributed (somewhat tentatively) to Perry. That argument relied on two main claims: firstly that if actions are to be caused by the contents of Cog, then actions must be correlated with changes in Cog (in response, in turn, to changes in relevant environmental factors); and secondly that only indexical belief states show the appropriate sensitivity to environment – non-indexical beliefs being too coarse-grained to explain many of the differences between the intentional behaviour of different agents, and of the same agent at different times.

Perry's most striking examples involve sensitivity to temporal changes, and to questions of an agent's own identity. In both cases it may be difficult to see how the sort of speculative counterexamples we used against the Targetting Argument are going to apply. Let us begin with times.

We described two ways in which non-demonstrative modes of presentation might target environmental features. One involved an extended symbolic language, capable of utilizing the outputs of perception without the need for demonstratively-grounded bridging conventions. The other involved an extended perceptual apparatus, capable of supplying a symbolic output of a kind presently utilized by Cog.

In a world in which times are as poorly distinguished as they are in ours, it is indeed difficult to see how times could be 'targetted' by a modification of the former kind. We do not appear to *perceive* times, as distinctive features of our perceptual fields. (Partly for this reason, it seems a mistake to try to treat temporal indexicals as demonstratives – to paraphrase 'now' as 'this time'. Temporal indexicals seem rather to rely on the fact that our perceptual experience – indeed our mental life as a whole – has a temporal parameter.) However, it is easy to imagine an environment that would allow cognitive systems to perceive times, in quite a natural sense: all that is necessary is that the environment contain some monotonically changing property, to which the cognitive systems concerned are perceptually sensitive. In a community of short-lived but sharp-eared creatures, a slowly changing musical tone might do the trick. So long as these creatures used mental representations of musical tones as cognitive symbols, it seems to me that they would not need temporal indexicals.

In any case, there are much more natural examples of the latter sort of modification, which consists in our perceptual apparatus being modified to suit our existing cognitive symbolism. Times have non-indexical names, provided by the standard calendar. Perry's examples show that for us, calendar dates and clock readings do not target the moments they refer to. Non-indexical temporal beliefs do not explain the fact that we act at one time rather than another. Conversely, we can target a time – intend to act at a particular moment – without knowing what calendar time it is. Even in the age of the digital watch, it would seem that most deliberately timed human actions are performed in ignorance of the information that such devices provide. This is hardly surprising, since our cognitive apparatus has presumably evolved in environments in which (the Great Watchmaker notwithstanding) useful timepieces were exceedingly rare. But the contingent constraints of past environments should not be confused for theoretical limitations. Whether robots or humans, we might in future do things differently. In particular, we might have accurate internal clocks, whose output would be constantly accessible to cognition. As in the pilot case, for example, this clock might be linked to an overlay on the visual field, so that at any time we would have direct visual access to a non-indexical symbol by means of which to refer to that time.

Imagine that we are built like this, and that we want to explain why I started reading at 11 o'clock. We mention my belief that the paper was scheduled to start at 11 o'clock, and my desire for punctuality. Is it necessary to add that at 11 o'clock I came to believe that it was then (i.e. 'now') 11 o'clock?

It seems to me not. However, there is an appealing argument to the contrary, stemming from the causal notion of explanation with which we are here concerned. (Indeed, it is an elaboration of the Causal Role Argument.) The function of indexical belief states is supposed to be to make the contents of Cog sensitive to a certain sort of environmental change: in the temporal case, either to changes in environmental indicators of time, or to 'changes' in time itself. (Here we can safely ignore the metaphysical problems associated with the notion of the passage of time.) The model is roughly this: an environmental change-in this case, a change in the time) is perceived by the agent, and registered as a change in an indexical belief state. (I come to believe that it is *now* 11 o'clock.) The newly-acquired indexical belief state then interacts with various non-indexical attitudes, and hence leads to an intentional action.

Thus the non-indexical part of Cog is being thought of as dispositional. It is *disposed* to respond in certain ways to changes in the indexical sector of Cog. The important question is whether what triggers the disposition has itself to be part of Cog. If so – and the Causal Role Argument depends on the claim that it is so – then Cog must contain the sort of state that is capable of exhibiting this kind of environmental sensitivity (and an overlay clock would not eliminate the need for temporal indexicals).

The principle that the triggers of action lie in Cog might be thought to follow from two independently plausible intuitions: first, that changes in effects require changes in causes; and secondly, that the environmental causes of intentional actions are mediated by the contents of Cog. Taken together, these intuitions may seem to require that relevant environmental changes must themselves be represented in Cog.

However, this argument is self-defeating; for there is an analogous argument involving the indexical and non-indexical parts of Cog. Changes in the indexical part (if any) of Cog are among the causes of intentional action. And their effects are mediated by the non-indexical part of Cog. But clearly we do not want to say that changes in the indexical part must be reflected in changes in the non-indexical part. The very purpose of the indexical part was to quarantine these changes.

To avoid this *reductio* of the indexical/non-indexical division of Cog, we presumably need to say something about the mediating role of the non-indexical part: for example that it acts like a conduit, conveying changes from the indexical part of Cog to the Will without itself changing in the process. Something like this will presumably make sense of the two intuitions we started with. But now the problem is obvious: why shouldn't we apply the conduit model in the first place, rather than introducing the indexical part of Cog? Why shouldn't we think of Cog as a whole as a conduit, conveying changes in our perceptual states to changes of the Will? The way thus seems clear for cases in which the non-indexical part of Cog responds directly to inputs from perception – inputs that are not themselves cognitive attitudes.

The overlay clock provides one such case. Equipped with such a clock, we would apparently have no use for perceptual reports such as 'It is now 11 o'clock'. This is not simply the pragmatic point, that it won't occur to any of us to doubt that I would have this belief at 11 o'clock (so that from a pragmatic point of view, the explanation of my starting to speak at 11 o'clock would not need to mention it). It is rather the argument that there is no longer a causal role for any such belief. The non-indexical attitudes I mentioned earlier (the belief that the paper was due to start at 11 o'clock, and the desire for punctuality) together constitute a disposition to respond to a certain change in my visual field: that change that takes place when my overlay clock flips over from 10:59 to 11:00.

It might be objected that clocks are never totally reliable, and that even if one of them was reliable, it would always be possible to doubt its reliability: to ask, when the clock shows 11:00, 'Is it really 11 o'clock'? And doesn't this question depend on the possibility of demonstrative reference to times? In other words, isn't what is being doubted here indexical in form? By itself this objection is not an argument for the indispensability of indexical belief states in action explanation. But once the possibility of doubt is admitted, it soon becomes one. For we then feel inclined to say that when I started reading at 11 o'clock, I must have trusted my internal clock – consciously or otherwise. I must have believed that it was *then* 11 o'clock.

However, this is just the Argument from Doubt. It fails here as it failed as an argument for Peacocke's IT. First, it depends on the assumption that belief is entailed by a lack of the corresponding disbelief. Secondly, and more importantly, it tacitly relies on the assumption that *only* indexical or demonstrative reference can make the contents of Cog causally sensitive to features of an agent's environment. But of course the example is intended precisely to illustrate how that assumption might fail, for a suitably equipped community of agents. (True, the example requires that such agents take for granted the reliability of their perceptual equipment. But in the previous section we observed that any perceptually-grounded system of reference requires something like this. In our case it is the tacit assumption that our perceptual apparatus is sufficiently reliable, and *shared*, for demonstrative or indexical reference to work at all.)

So much for times. However, Perry's most striking examples concern an agent's awareness of his own identity. Perry imagines himself in a supermarket, spilling sugar from a broken packet. Seeing a reflection in a mirror, he may believe that *that* man is making a mess. He may even believe that John Perry is making a mess. But it is only when he comes to believe that he *himself* is making a mess (i.e. what he would express by 'I am making a mess') that he is moved to do something about the spilling sugar.

Dramatic as these sorts of examples are, it seems to me that there are no problems here not already encountered in the temporal case. If anything, personal indexicals are easier to eliminate than temporal ones; for self-identities, unlike times, are cognitive constants. Barring forgetfulness, knowledge of one's own identity need be gained only once. It follows that in principle perception need have nothing to do with the acquisition of this sort of knowledge: we could simply be made with it. In other words, each of us could be pre-programmed to use a personal name, unique to ourselves, as we now learn to use the indexical 'I'.

Others could use our personal names, for example to convey information about us to ourselves and third parties. Huw Price could thus inform John Perry that John Perry is making a mess. Both of us would then believe what we would both express by saying 'John Perry is making a mess'. The rest of our relevant beliefs and desires may also match, but Perry acts and Price does not. How can this be, if both are cognitive systems, acting on the basis of their beliefs and desires? In virtue of the fact that in one crucial respect, the same cognitive systems are wired into different agents. Perry is wired to act on the basis of beliefs of the form *Perry should do 'X'* (or desires of the form *Pery wants to do 'X'*), whereas Price is programmed to act on the analogous beliefs or desires about Price.

It might be objected that the example actually supports Perry's case, in that it illustrates that non-indexical beliefs are not in general sufficient to explain action (since in such a case we have to refer to the sort of agent in which the beliefs and desires are embodied). However, I agree that non-indexical beliefs do not suffice to explain action. I am only denying that this entails that intentional agents must have indexical beliefs. The point of this example, as of the others, is simply to show that the explanatory work normally done by indexical (or demonstrative) beliefs can be done in other ways. Suitably designed cognitive systems would not need such 'locating beliefs' (as Perry calls them). Instead, the task of 'locating' their entirely non-indexical Cogs at a particular point in their environment would be handled by their perceptual and motor components.

I conclude that in so far as Perry's argument differs from Peacocke's, it is no more successful. In particular, the Causal Role Argument is invalid. Equipped with appropriate ancillaries, an agent can be sensitive to environmental changes, without representing in Cog the environmental differences which are relevant to action explanation. Indexicals, like demonstratives, are essential only for creatures of our present and thoroughly contingent limitations – and even then not all the time. As for robots, they would be well advised to help themselves to in-built clocks and pre-programmed personal names (and to on-board position sensors, for that matter); and hence – among themselves, at any rate – to escape indexical complexities.[14]

5. THE END OF DESIRE?

What does Hume's argument have in common with those of Peacocke and Perry? Obviously all three appeal to action explanation, but is there any more to it than that? In this final section, I shall try to show that there is. Hume relies on the principle that beliefs are motivationally inert. I shall argue that this principle's plausibility derives from arguments and assumptions that run parallel to some of those we have seen to underlie the arguments of Peacocke and Perry (and hence that it is vulnerable to similar objections).

First an objection: Hume's principle might seem to be entailed by a perfectly orthodox view of the explanation of action. Almost everybody agrees that at least one way to explain an action is describe the agent's *reasons* for doing it – the beliefs and desires that led him or her to act in that particular way. In challenging Hume's argument, wouldn't we be challenging this whole picture of action explanation?

To the extent that this view of action explanation derives from the Humean model, I think it is true that in criticizing Hume's principle we shall be offering reasons for a critical reassessment of the orthodox model. However, for present purposes we need not argue for the rejection of this model. For in one crucial respect Hume's conclusion is stronger than the model requires. It would be quite in keeping with the orthodox view to say that the explanation of action involves two sorts of belief: beliefs about the desirability, or value, of certain states of affairs; and beliefs about the bearing of the action in question on the obtaining of those states of affairs. Hume's principle, on the other hand, claims that belief alone cannot motivate action – in other words, that the explanation of action requires mention of states of mind *that are not beliefs*.

This difference is crucial to the two applications I mentioned in section 1. On the one hand, Hume's principle is of interest to cognitive theorists precisely because it demands a category of cognitive states distinct from beliefs. If true, it is thus a major constraint on the large scale structure of cognitive systems. It implies that agents must embody at least two different 'boxes' of cognitive attitudes, and hence that as theorists we need at least two distinct accounts of the relation of the content of an attitude to its cognitive role. If desires could be regarded as a kind of belief, things would seem much simpler. The task of explaining the distinctive cognitive role of a desire would then fall to the general theory of content (which has lots of explaining to do anyway), and we could hope for a unitary account of the relation of content to cognitive role.

Similarly, Hume's distinction between motivating desires and non-motivating beliefs is crucial to the use that is made of Hume's argument by non-cognitivists in ethics and aesthetics. Non-cognitivists want to show for example that moral judgements are not *factual* judgements in the ordinary sense. This requires a general distinction between fact-stating and non-fact-stating uses of language. Typically, at any rate, non-cognitivists rest this *semantic* distinction on a *psychological* one: genuine *statements of fact* are taken to be the expressions of genuine *beliefs*.

The non-cognitivists then want to show that moral statements lie on the right side of this psychologically-grounded distinction – in other words, that such statements do not express beliefs. Given Hume's principle, it is enough to show that moral commitments do motivate actions; for this puts them on the desire side of Hume's distinction. And to many, at any rate, it "seems to be a conceptual truth that to regard something as good is to feel a pull towards promoting or choosing it".[15] Clearly, someone who rejected Hume's principle could quite well endorse the standard model of action explanation, and accept that moral judgements express mental states on the active or motivating side, and yet insist that moral commitments are genuinely beliefs about moral facts.

Note, however, that such a view admits two possible accounts of the place of desires. One account would recognize a category of desires, distinct from beliefs; but claim that motivation is not the exclusive preserve of attitudes of this category (moral commitments being one sort of exception). The problem will then be to say what distinguishes beliefs from desires, given that it is not the power to motivate action. This problem might incline us to the second sort of account, which would recognize no useful distinction between beliefs and desires, claiming that all propositional attitudes can be represented as species of belief.

These two accounts agree in rejecting Hume's principle – in other words, in accepting the possibility of motivating beliefs. Why, these days (freed of the other peculiarities of Hume's philosophy of mind), should anyone deny this possibility? The first line of argument, it seems to me, is an analogue of the Argument from Doubt. We might call it the *Argument from Dislike*: for any property *F* of an action or state of affairs, it is conceivable that an agent could be motivated by a desire that not *F*. This being so (the argument runs) no action can be explained simply in terms of a belief that the action concerned would be *F* (or would produce some *F* state of affairs). We need to add that the agent desired that *F*, or at least did not desire that not *F*. In other words we need to mention the agent's desires.

I think it should be clear that this argument depends on the same kind of assumption as the Argument from Doubt: that the fact that an explanation would be inadequate in the presence of some extra condition, entails that it is inadequate unless it mentions the absence of that further condition. The assumption is no more plausible here than it was for demonstrative beliefs (or for causal explanations in general). As in the demonstrative case, to admit the assumption here would be to abandon any systematic approach to the contents of Cog. For there is no end to the things that even the most fastidious agent does not dislike.

From the point of view of the second of the two accounts we recently distinguished, there is another objection to the Argument from Dislike. For this account sees the question at issue not as whether agents have desires, but as whether these desires are anything but a special class of beliefs. On the latter view – the view the argument is intended to refute – to believe that an outcome of a possible action is *F* might be the very same thing as desiring that outcome. In this case the possibility that the argument imagines – that of someone holding the belief, but desiring an outcome such that not *F* – would be that of a person who desires what he takes to be undesirable. We seem entitled to regard this as no possibility at all.

At the corresponding point in our discussion of Peacocke's argument, we asked why doubt (in demonstrative form) should matter. If an agent wants the bottle containing the pills, why should it matter if he doubts whether *that* bottle contains the pills, more than it matters if he doubts whether the bottle that belonged to Uncle Fred contains the pills? The answer seemed to lie in the assumption that in virtue of its special association with perception, the demonstrative thought plays a crucial role in enabling an action to connect with its target.

Is there an analogous question in the present case? I think that there is. The argument again turns on the connection between two mental components of a fully-fledged intentional agent: on the one hand Cog, the seat of cognitive deliberation; on the other, a unit whose function is to initiate bodily action, in response to the deliberations of Cog. There are a number of possible models of this latter unit – let us call it 'the Will' – and of its relation to Cog. At one extreme we could regard it as a kind of store, or processing house, for intentions already formed in Cog. At the other extreme we could take it to be the factory in which judgements arriving from Cog are matched with a store of desires to yield intentions. However, since we want a model in which the action which interests us takes place at the boundary between Cog and the Will, let us avoid both extremes. Let us view the Will as the surface on which intentions 'condense', as a consequence of reactions that take place in Cog.

In effect, Hume's claim is then that no intention can condense in the absence of a desire. The Argument from Dislike claims that whatever beliefs are present, we can imagine a desire whose presence would prevent an intention condensing from those beliefs. Its rebuttal points out, first, that even if this were true, it might well be irrelevant to the question at issue, which is whether in the absence of that desire (or any other), these beliefs are causally sufficient for the formation of the given intention; and secondly that if we are allowing the possibility of desires that are themselves beliefs, then in the crucial cases the claim simply isn't true. Some beliefs are not compatible with extra desires.

All the same, it may be felt that neither reply tackles the intuition which – perhaps mistakenly – is taken to be validated by the Argument from Dislike. Like the Argument from Doubt, this argument may be a blind alley. But why should it seem appealing in the first place? What is special about desires, in virtue of which they are at least capable of pre-empting other motivations?

In section 3 the analogous question about demonstratives led us to the Targetting Argument – to the idea that in virtue of their perceptual basis, demonstratives play a crucial role in cognitive control of perceptu ally-guided action. In the present case I think the answer alsoturns on a special causal role. Desires are being held to fill a causal and explana tory lacuna between beliefs on the one hand, and the Will, on the other.

Desires are the catalysts that cause intentions to condense from beliefs. The Argument from Dislike attempts to establish the need for such a catalyst, by claiming that in conceivable circumstances, a given set of beliefs would not give rise to the formation of a certain intention. We know that the argument fails. But the catalytic view of desire has a lingering appeal.

At this point in the demonstrative and indexical cases, we looked for actual and hypothetical examples to show that the action-guiding causal role of demonstrative modes of presentation could be performed by non-demonstratives. We tried to show that that role is not a kind of cognitive constant, which would have to be tackled in the same way in any conceivable cognitive system. On the contrary, its nature depends very much on contingent features of the system's environment, its perceptual apparatus, and the symbolic resources of its cognitive centre.

My suggestion now – not a very surprising one, I think – is to adopt the same strategy in the present case, against the catalytic model of desire. That model sees a causal gap between beliefs, on the one hand, and the Will, on the other: intentions cannot condense in the absence of desires. My point is that whether this is so depends on the nature first of beliefs themselves, and secondly of the surface on which intentions are going to form – that is, of the Will. There is no reason to regard these factors as cognitive constants. On the contrary, we shall see that it is easy to imagine agents who differ in these respects, from each other, and from ourselves. In consequence, a non-catalytic intention condenser is far from inconceivable. The interesting question is whether there actually are any.

As in the demonstrative case, there are two main possibilities: where there is a causal gap, it can be closed from either side. (Again I shall ignore the obvious hybrid possibilities, and also solutions involving a separate bridging component.) The first option is to expand an agent's belief system, so as to admit beliefs that would give rise to intentions directly (or at least as directly as belief-desire combinations are ordinarily held to do so). We want a belief which would *in itself* normally comprise a reason for performing an action – a belief whose capacity to rationalize an action does not depend on the co-occurrence of certain desires.

I think we can get a good idea of what such a belief would be like, by reflecting on familiar expressions of moral or rational necessity. For example: *I must go down to the sea again*; or *I am bound to tell the truth*.

Superficially, at any rate, utterances such as these can be regarded as expressions of a belief that a possible action (going down to the sea, or telling the truth) has a certain property (being what I must do, or am bound to do). Moreover, these beliefs (if they really are beliefs) seem to provide reasons for actions that are insensitive to an agent's desires about actions with the corresponding properties. If I believe that I must go down to the sea again, then it seems I have a reason for going down to the sea even if I do not particularly want to do what I must do. (A man must do what a man must do, whether he wants to or not.) Similarly, if I think I am bound to tell the truth, it seems irrelevant whether I want to do so.

These points are (notoriously) disputable. There are two lines of attack. One urges that believing that I must do something, or am bound to do it, only provides me with a reason for doing it because, like everybody else, I have an implicit desire to do what I must. The other takes the opposite tack, and argues that because utterances such as these express reasons for actions, they do not express beliefs (and hence have to be construed some other way). However, both lines of argument seem to depend on the very claim at issue: that beliefs alone cannot motivate actions.

It is indeed contentious whether these expressions of felt necessity do, as their form suggests, express genuine beliefs. To tackle the question, however, we need a criterion with which to distinguish beliefs from desires. Hume's principle offers such a criterion, saying that desires are distinguished in virtue of their active or motivating role in the production of action. But we now find that the principle itself depends on the criterion it claims to supply.[16]

Thus it seems that we have no non-circular reason to deny the possibility of an *intrinsically rationalizing belief*: a belief about a possible action whose capacity to rationalize that action is independent of an agent's desires. And if we have no reason actually to affirm the possibility, this is because the distinction between belief and desire is itself so obscure (one reason, I think, why the argument benefits from comparison to the relatively clear demonstrative and indexical cases).

I said that as in the demonstrative case, there are two main ways to bridge the gap that Hume's principle claims to find between beliefs and the Will. The second is to imagine the Will itself equipped to handle a somewhat broader range of inputs (just as we had to think of an agent's perceptual system as producing a broader range of outputs). Why shouldn't the Will, or the unit in which an agent forms intentions to act, simply be built to respond to certain beliefs about the properties of actions and states of affairs? A familiar example is the classical economic rationalist, whose actions are based entirely on the profit motive: she forms the intention to do whatever she believes will maximize her net monetary return. It might be objected that the rationalist *wants* to make a profit. But why should she? If she has desires at all, why shouldn't she be built so that desiring something and thinking it profitable are one and the same?

A similar case is the perfect (reflective) Bayesian, who chooses the action he takes to maximize his expected utility. Whether he has desires concerning possible outcomes or beliefs concerning their utilities, his final intention to act must be based simply on the *belief* that that action maximizes expected utility. For if we have to add that he *wants* to maximize expected utility, then in order to do so he has to make a further calculation, taking that want into account. To insist on a catalytic desire would be to prevent him from acting at all.

These cases are admittedly unrealistic. But they show, I think, that there is no conceptual obstacle to the idea of an agent who is motivated to act by beliefs that in our own case would need to be supplemented by other attitudes. We are not economic rationalists or perfect reflective Bayesians. But we can imagine circumstances in which our deliberations and choices would parallel those of such agents. In the rationalist case, for example, we can see that this would happen if we acquired an overriding desire for monetary gain. Modulo that desire, we remain deliberative agents; and hence it is a short step to a conception of an agent in whom the function of that desire is factored out – in whom the Will, the intention-forming psychological component, is simply structured to respond to beliefs of the appropriate sort.

Thus it seems to me that the case for Hume's principle is indeed vulnerable to parallel objections to those we have raised against the arguments of Peacocke and Perry. It derives some unwarranted plausibility from an analogue of the Argument from Doubt. And more importantly, it turns out to depend on unjustified assumptions about the causal connections between the mental 'organs' of a cognitive agent.

The major difference, I think, concerns the bearing of these conclusions on the respective applications of the two sorts of argument. In the indexical case, the conclusion seemed of largely theoretical interest.

There is no doubt that in practice we do depend heavily on demonstrative and indexical modes of presentation of features of our environment. The mistake is to try to generalize on this, extrapolating to arbitrary cognitive systems what are really peculiarities of our own situation.

In the desire case, however, the conclusion lies closer to home. In this case the practical question not only remains unsettled, in the sense that is is not clear to what extent our actions are in fact explained by attitudes that are not beliefs; but also the question itself is under threat. Hume's principle is not only taken to show that action required desires, as well as beliefs; it has also provided a reason for *distinguishing* desires from beliefs. Without it, we lack a foundation for that distinction. As cognitive scientists, this means that we lack a justification for the traditional two-box taxonomy of cognitive states – a reason not to opt for a single species of propositional attitude, differentiated entirely by *content*. While as non-cognitivists, it means that having sought to ground our semantic distinction psychologically, we may be forced to retrace our steps. Like Hume himself, we may feel bound to say that desires are distinguished from beliefs in virtue of the fact that it is "impossible (. . .) they can be pronounced either true or false".[17] (Of course, the problem then will be to show that moral commitments are not true or false.)

In either case, the prospect may be so uninviting, or at least so unconventional, that we choose to defend Hume's conclusion. The point of this excursion has been to show that its present defences are rather inadequate. Like many a tradition, it survives at present not so much on established merits as simply in virtue of not being challenged. But nothing challenges tradition like the discovery that others might do things differently. Hence this brief excursion to the possible worlds. In order to counter the tendency to unwarranted generalization from our own case, I have tried to encourage a flexible and speculative approach to the cognitive functions that have seemed to necessitate certain sorts of mental states. What travel broadens, in this case, is one's view of the possible structure of the mind.

NOTES

* I am grateful for comments from Hugh Mellor, Philip Pettit, Phillip Staines and participants in a seminar at ANU in 1986; also for financial support from a National Research Fellowship.

1 David Hume, *Treatise*, Book II. Part III, Section III.

[2] Following current convention, I shall use the term 'desire'. I do not, of course, mean to claim that in a proper taxonomy of motivating states of mind, 'desire' would be an appropriate general term.
[3] *Synthese* **49** (1981), pp. 187–217. (Hereafter, 'DTPE'.)
[4] DTPE, pp. 205–206.
[5] Since I cannot see what other arguments Peacocke could have in mind, I assume that he himself regards IT as having this sort of generality. But my argument is not *ad hominem*: the important question here is whether IT bears this generalization, not whether Peacocke wanted it to.
[6] *Nous* **13** (1979), pp. 3–21. (Hereafter, 'PEI'.)
[7] As in Peacocke's case, however, my argument is not *ad hominem*, and the reader should feel free to regard my Perry as a straw man.
[8] PEI, p. 4.
[9] DTPE, p. 207.
[10] DTPE, p. 210.
[11] Not as non-specific as an intentional action can be, however. Intelligent spiders might still spin webs. And many of *our* intentional actions are quite as non-specific as the spider's fly-catching. Draw a breath, and you'll see what I mean.
[12] This view of the relationship between names and objects is relatively modern, however. The medieval view seems to have been that before Babylon, at any rate, names were iconic – based on similarities between the names themselves and what they represented (see, e.g., Michel Foucault (1970). *The Order of Things*, Tavistock Publications, chapter 2). If so, then at one time it would have been much more plausible to say that physical objects were targetted by their non-demonstrative names. If names were iconic, then Cog could present the perceptual guidance system with a picture of its target.
[13] Wittgenstein comments on this sort of possibility, relating it to our use of words as names of themselves: "What about (. . .) colour-samples (. . .): are they part of the *language*? Well, it's as you please. They do not belong among the words; yet when I say to someone: 'Pronounce the word "the"' you will count the second 'the' as part of the sentence. But it has a role just like that of a colour-sample (. . .); that is, it is a sample of what the other is meant to say". (*Investigations*, 16)
[14] I emphasize, however, that this conclusion lies firmly in the realm of what I earlier called 'speculative psychology', and therefore seems compatible with Perry's main conclusion. Perry wants to argue "that the essential indexical poses a problem for various otherwise plausible accounts of belief" (PEI, p. 3). Obviously it is no defence of these accounts to point out that we would not need the indexical, if only we were built differently.
[15] Simon Blackburn, *Spreading the Word*, Oxford University Press, 1984, p. 188.
[16] As McDowell (1978) puts it, "the idea of the world as motivationally inert is not an independent hard datum. It is simply the metaphysical counterpart of the thesis that states of will and cognitive states are distinct existences; which is exactly what is in question". ('Are moral requirements hypothetical imperatives?', *Proceedings of the Aristotelian Society, Supplementary Volume* **52** pp. 13–29; at p. 19.)
[17] *Treatise*, Book **III**, Part I, Section I.

Department of Philosophy, Research School of Social Sciences
Australian National University, Canberra, Australia

INDEX OF NAMES